Multi-objective Design Optimization of Switched Reluctance Motor Drive Systems

Xiaodong Sun · Kaikai Diao · Gang Lei · Jianguo Zhu

Multi-objective Design Optimization of Switched Reluctance Motor Drive Systems

Springer

Xiaodong Sun
Automotive Engineering Research Institute
Jiangsu University
Zhenjiang, China

Kaikai Diao
School of Electrical Engineering and Automation
Anhui University
Hefei, China

Gang Lei
School of Electrical and Data Engineering
University of Technology Sydney
Ultimo, NSW, Australia

Jianguo Zhu
School of Electrical and Computer Engineering
University of Sydney
Darlington, NSW, Australia

ISBN 978-981-96-0671-9 ISBN 978-981-96-0672-6 (eBook)
https://doi.org/10.1007/978-981-96-0672-6

This Springer imprint is published by the registered company Springer Nature Singapore Pte Ltd.
The registered company address is: 152 Beach Road, #21-01/04 Gateway East, Singapore 189721, Singapore

Contents

Chapter 1
Introduction

Abstract This chapter briefly introduces switched reluctance motor (SRM) drive systems and their state-of-the-art design optimization methods. The design optimization of SRM drive systems is a system-level, multi-objective, highly coupled, and nonlinear problem, which makes it a significant challenge to achieve the optimal solution. This chapter provides readers with a basic understanding of SRMs. It paves a smooth path to present various optimization methods for achieving optimal results according to the design specifications of many challenging applications, such as aircraft and electric vehicles.

Keywords Design method · Drive systems · Electrification · Multi-objective optimization · Switched reluctance motors

1.1 Energy Challenges and Electrification Development

Energy challenges and electrification development are crucial topics worldwide, as the energy demand continues to rise while conventional energy sources become scarce and expensive. The transition towards sustainable energy systems is becoming increasingly urgent, and electrification is vital in achieving this transition. Electrification refers to adopting electric power as a substitute for conventional forms of energy. It can be implemented through the widespread deployment of renewable energy technologies and the electrification of transportation and industries [1, 2].

Figure 1.1 shows the proportion of different types of energy resources used for electricity generation from 2010 to 2021 globally [3]. The data is based on information from the International Energy Agency (IEA) and shows that coal has been the most commonly used energy resource for electricity generation. Hydroelectricity has consistently been the third most used energy resource, with a proportion ranging from about 16% in 2010 to 18% in 2021. Nuclear energy has been used consistently. The use of oil for electricity generation has decreased over time, with a proportion ranging from nearly 6% in 2010 to 3% in 2021.

X. Sun et al., *Multi-objective Design Optimization of Switched Reluctance Motor Drive Systems*, https://doi.org/10.1007/978-981-96-0672-6_1

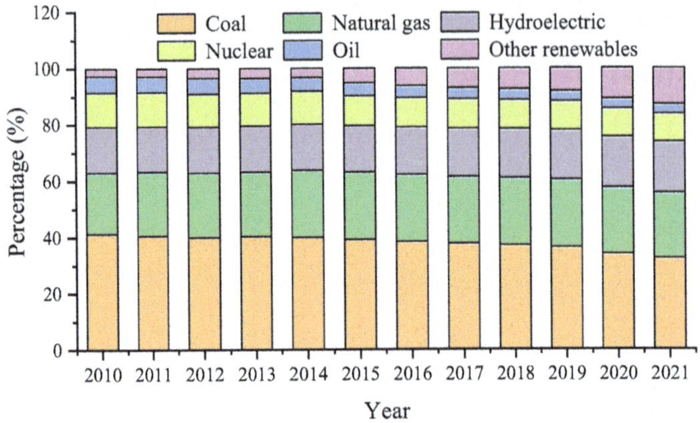

Fig. 1.1 The data source for the proportion of different types of energy resources used for electricity generation from 2010 to 2021 [3]

Moreover, the use of other renewables except hydroelectricity has increased by about 10% in 2021. The increasing use of renewables for electricity generation is a positive trend, likely driven by the decreasing costs of renewable energy technologies and growing awareness of the environmental impacts of using fossil fuels. However, the use of fossil fuels for electricity generation remains dominant globally, highlighting the need for continued efforts to transition towards cleaner and more sustainable energy sources.

On the other hand, the data on global annual mean surface temperature and average carbon dioxide (CO_2) concentrations from 1900 highlight the urgent need for a transition to renewable energy sources, as shown in Fig. 1.2; [4, 5]. Burning fossil fuels is the main contributor to the increase in CO_2 concentrations, which drives global warming and climate change. The link between the two datasets also shows that the rise in surface temperature is not a one-off event but a long-term trend that requires sustained and coordinated action. The shift to renewable energy must be accompanied by efforts to improve energy efficiency, reduce waste, and increase the use of low-carbon transportation modes.

According to the IEA's 2021 report on global energy review [6], the global electricity demand is distributed across various sectors, with the buildings and industry sectors alone accounting for more than 70% of the total demand, as shown in Fig. 1.3. Additionally, the transportation sector accounts for a significant portion of the demand, highlighting the importance of electrifying transportation to reduce greenhouse gas emissions and improve air quality. The portion of electrical energy is expected to increase in the transportation sector with the penetration of more efficient electrified vehicles, including hybrid, plug-in hybrid, and battery electric vehicles (EVs).

In terms of equipment types, as shown in Fig. 1.4, motors are the most significant contributor to global electricity consumption, accounting for approximately 45%

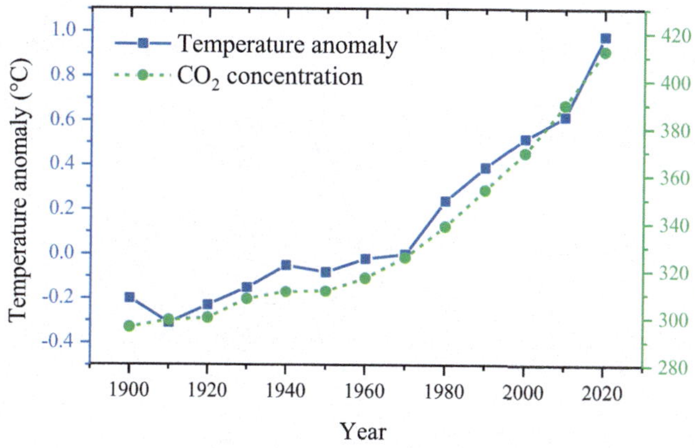

Fig. 1.2 The data source for global annual mean surface temperature and global annual average CO_2 concentration since 1900 [4, 5]

Fig. 1.3 Global electricity demand by end-use sectors [6]

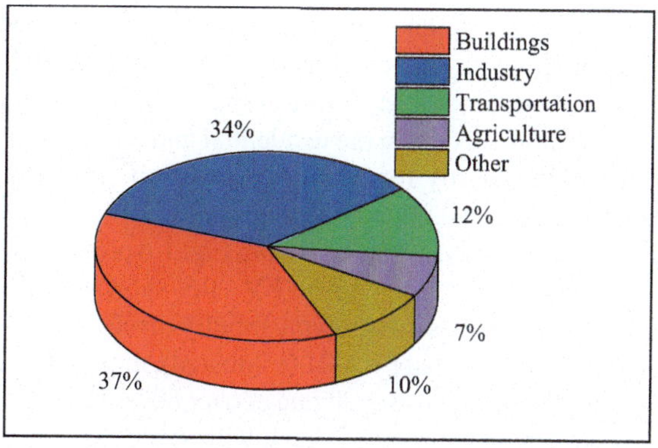

of the total, followed by lighting at 15%, heating, ventilation, and air conditioning (HVAC) systems at 10%, and information and communication technology (ICT) devices at 7%. Appliances and other types of equipment account for the remaining 23% of global electricity consumption. It is important to note that these figures are approximate and can vary depending on various factors, such as geographic region, energy efficiency policies, and technological advancements. Nonetheless, the data provides valuable insight into the primary contributors to global electricity consumption and can be used to inform policy decisions to reduce energy consumption and promote sustainability.

This data emphasizes the importance of improving the energy efficiency of electric motor systems in industries and other sectors where electric motors are widely used. Most electric motor energy consumption occurs in industrial processes, where

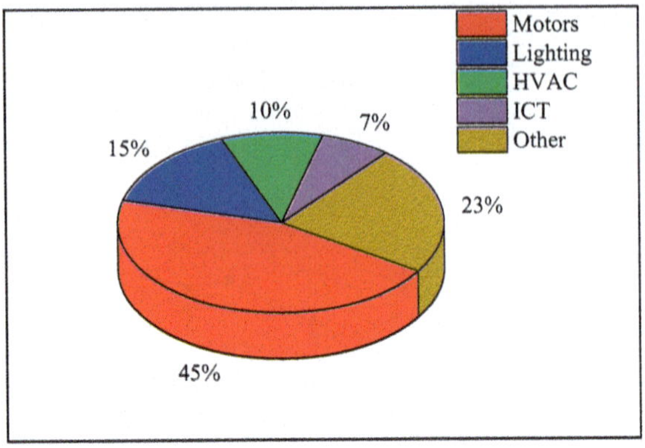

Fig. 1.4 Global electricity demand by equipment types [6]

electric motors are used in various applications, such as pumps, fans, compressors, conveyors, and mixers. Among electric motors used in industrial applications, the most common applications are pumps, which account for about 31% of energy consumption, followed by fans (20%) and compressors (19%). Other industrial applications include conveyors (8%), mixers (5%), extruders (3%), and other applications (14%), as shown in Fig. 1.5; [7]. The remaining electric motor energy consumption occurs in the residential and commercial sectors, where electric motors are used primarily for heating, ventilation, and air conditioning (HVAC) systems, refrigeration, and other appliances.

As for the transportation sector, Fig. 1.6 presents the data for global EV sales from 2010 to 2021 [8]. The data on EV sales show a significant demand increase over the past decade, with sales growing rapidly in recent years. In 2010, global EV sales were just 0.01 million, but by 2021 (estimated), they had reached 4.42 million, representing a compound annual growth rate of over 60%. The growth in the EV market is expected to continue in the coming years, with some analysts predicting that EVs could make up as much as 50% of global passenger vehicle sales by 2030. This shift towards

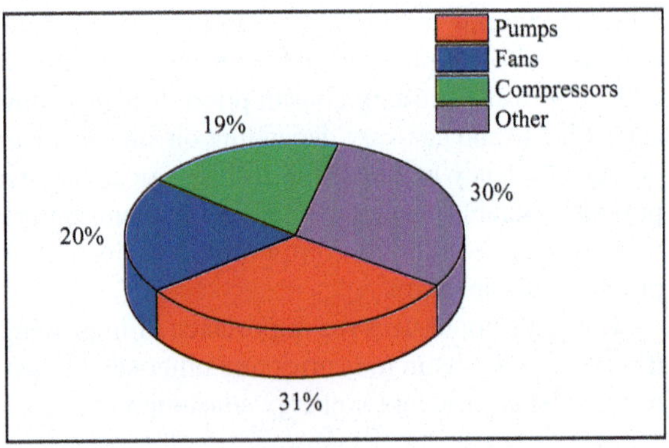

Fig. 1.5 Electric motors used in industrial applications [7]

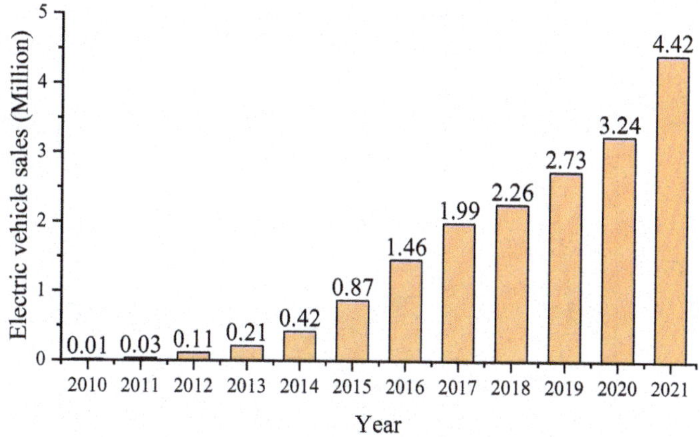

Fig. 1.6 Global EV sales from 2010 to 2021 [8]

electrification is also expected to drive demand for electrical machines, such as electric motors and power electronics, which are critical components of EVs. The bright future of EVs and electrical machines is driven by their benefits over conventional gasoline-powered vehicles, including lower emissions, reduced dependence on fossil fuels, and lower operating costs [9, 10]. As governments worldwide set ambitious targets for reducing greenhouse gas emissions, the transition to EVs and electrical machines will likely accelerate, creating opportunities for innovation, investment, and job creation in the clean energy sector.

Overall, the data on global electricity demand by sector and electric motors' significant role in global electricity consumption highlight the need for comprehensive and integrated systems to address energy challenges and electrification development. Such systems should focus on improving energy efficiency and promoting the adoption of renewable energy technologies in various sectors and end-uses, including electric motors in industries and transportation.

1.2 Brief Introduction of Switched Reluctance Motors

Permanent magnet (PM) motors have become increasingly popular due to their high efficiency, reliability, and low maintenance requirements [11]. However, one major limitation of PM motors is the price of the rare-earth materials and magnets. These materials, such as neodymium and dysprosium, are relatively scarce and expensive, which can drive up the cost of the motor [12].

The analysis of the United States Geological Survey (USGS) revealed several interesting trends and patterns in the pricing of neodymium and dysprosium, as shown in Fig. 1.7. For neodymium, the price per kilogram reached its peak in 2011 at $459/kg, followed by a significant drop in the following years due to an oversupply

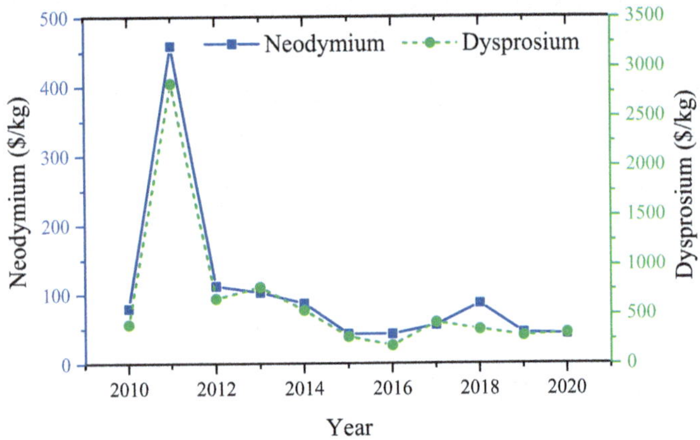

Fig. 1.7 Price of neodymium and dysprosium between 2010 and 2020 [13]

in the market. In 2015 and 2016, the price remained stable at around $43/kg before slightly increasing in 2017 and 2018. However, in 2019 and 2020, the price dropped again to around $45/kg and $43/kg, respectively. In contrast, the pricing of dysprosium showed a more consistent trend during the analysis period. The price per kilogram peaked in 2011 at $2835/kg before gradually declining in the following years. However, even with this decline, the price of dysprosium remained high compared to neodymium, with an average price of around $300/kg in recent years [13].

The analysis of the USGS data suggests that the price of neodymium and dysprosium has fluctuated significantly over the years, influenced by market forces and geopolitical factors. These fluctuations affect the cost of PM motors, which rely heavily on these materials. As the demand for EVs and other high-tech applications continues to grow, it is crucial to consider the potential impact of these price fluctuations on the development and adoption of sustainable technologies.

To address this issue, many researchers and engineers have begun exploring alternative motor designs that do not rely on PMs. One such design is the SRM, which uses a simple rotor made of iron or other low-cost materials and a stator with a series of coils. Several conventional topologies of SRMs are presented in Fig. 1.8. The operation of an SRM is based on the principle of magnetic reluctance, which is the tendency of a magnetic circuit to resist the flow of magnetic flux. By controlling the timing and duration of the current flowing through the stator coils, the motor can generate a rotating magnetic field that interacts with the rotor to produce torque. SRMs offer several advantages over PM motors, including lower cost, greater durability, and better high-temperature performance [14]. They are also more convenient to manufacture and maintain since they do not require PMs [15, 16].

In addition, SRMs have high torque density, which makes them suitable for high-performance applications, such as EVs, aerospace, robotics, and renewable energy systems. Despite the advantages of SRMs, they also face challenges, such as big torque ripple and acoustic noise, which can affect their performance and efficiency

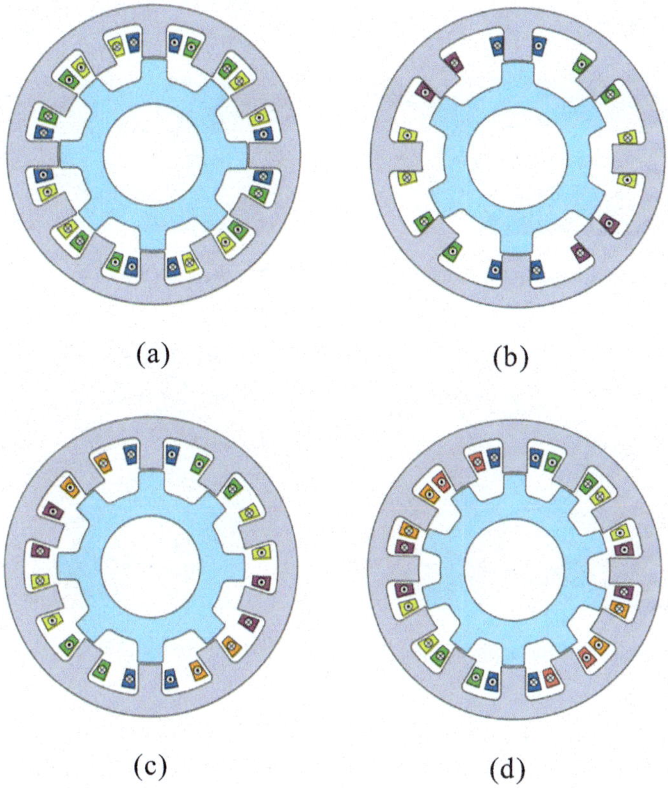

Fig. 1.8 Topologies of SRMs, a three-phase 12/8 SRM, b four-phase 8/6 SRM, c five-phase 10/8 SRM, and d six-phase 12/10 SRM

[17–20]. Researchers have recently proposed various techniques to address these issues, such as sensorless control and optimization algorithms [21–26].

From the view of EV application, many researchers have made preliminary attempts to use SRMs as drive motors. The most representative work is that Chiba et al. designed a 50 kW SRM by matching the dimensions and various performance indicators of the second-generation Toyota Prius motor, as shown in Fig. 1.9; [27–32]. A series of design optimization and measurement research have been carried out. The 0.1 mm 10JNEX900 silicon steel sheet is selected to improve the overall efficiency, and the continuous current operation control is used to broaden the constant power range of the motor [27–29]. The designed motor can reach the torque, efficiency, and speed regulation range of the second-generation Prius motor, and its highest efficiency can reach 95% [30–32]. With the advent of the third-generation Toyota Prius motor, the team further designed a matching 60 kW SRM. The actual measurement shows that in the range of 5400–13,900 rev/min, the SRM has high performance. The power density can be increased from 60 to 100 kW [33].

Fig. 1.9 Research on the application of SRM in centralized drive EV [27–32]

Fig. 1.10 Research on the application of SRM in distributed drive EV [34]

Another example is the double-stator axial flux SRM developed for distributed drive EV [34]. The vehicle and motor installation positions are shown in Fig. 1.10. Compared with the 2 kW PM motor of the same specifications, the axial flux SRM exhibits better torque characteristics. More examples of SRMs applied in EVs, aircraft and aerospace systems, wind generators, high-speed drives, and energy storage systems can be found in [35].

In conclusion, while PM motors have been popular in many applications, the rising cost of rare-earth materials has increased the interest in alternative motor designs such as SRMs. The application of SRMs has gained considerable attention in various fields due to their unique features and advantages. EVs, aerospace, robotics, and renewable energy systems are among the many applications where SRMs have shown potential. However, challenges such as high torque ripple and acoustic noise must be addressed to fully realize the potential of SRMs in these applications.

1.3 The State-of-Art Design Optimization Methods

Design and optimization are two major coupled stages when developing electrical machines. The primary purpose of the design stage is to find a feasible solution meeting the specifications by investigating the machine type/topology, materials, and dimensions. Optimization of electrical machines aims to identify the best configuration(s) among a great number of evaluated design candidates [36–40]. With the development of metamodeling techniques and increased computing power, most electrical machine optimization problems are performed using intelligent algorithms. Several innovative multi-objective algorithms have been proposed and applied to machine optimization, such as the multi-objective genetic algorithm [41–43], multi-objective particle swarm optimization algorithm [44, 45], and multi-objective differential evolution optimization algorithm [46, 47].

The design methods of SRMs can be divided into two main categories, i.e., the analytical and numerical methods [15]. Analytical methods involve solving the mathematical equations that describe the motor's performance, which is generally obtained based on Maxwell equations or the magnetic circuit models. Numerical methods, such as finite element analysis (FEA), are widely used to simulate the motor's magnetic field and optimize the electromagnetic design. FEA can provide accurate predictions of the motor's performance and can be used to optimize the design parameters [48].

In recent years, a variety of optimization methods have been proposed for SRMs. As reported in [49], a multi-objective design optimization was proposed based on the design of experiments and particle swarm optimization. In [48], a system-level optimization method has been investigated for the SRM drive systems, and an exemplary problem consisting of an SRM and its angle position control was implemented to achieve the best solution for the system. Some popular solutions like sensitivity analysis [16, 50, 51], surrogate modeling techniques [52–54], multi-level optimization approach [43, 55], and the genetic algorithm [56, 57] could be employed in the optimization process to improve the optimization efficiency. Besides, more studies investigated robust optimization methods that take the inevitable manufacturing imperfections and assembly errors into account, including the Taguchi method [58], worst-case design [59], and design for Six Sigma [60]. For instance, a system-level sequential Taguchi method was presented in [58] for the robust optimization of SRM, and the proposed method can eliminate the conventional Taguchi method's dependence on the defined levels.

1.4 Primary Objectives of the Book

This book presents multi-objective design and optimization approaches for advanced SRM drive systems. The design optimization methods mainly include the working principle, modeling, control, topology design, and optimization techniques. This book will benefit researchers and engineers designing and optimizing SRM drive systems and other types of motors. The methods proposed in this book could guide the design and optimization of electro-mechanical systems. The main objectives are listed as follows:

(1) To present the design and optimization foundations of SRMs, including the working principle, modeling methods, control methods, novel topologies, and optimization classification. These are prerequisites for further investigation.
(2) To introduce the conventional design methods and develop a novel magnetic parameter design method based on the nonlinear characteristic of SRMs.
(3) To review the popular optimization techniques for electrical machines, including the design of experiments, sensitivity analysis methods, surrogate models, and optimization algorithms, and present examples for applying them to SRMs' optimization.
(4) To develop the multi-objective deterministic optimization method for SRM drive systems to realize the tradeoff between different optimization objectives, which will consider both motor structure and control aspects.
(5) To develop a multi-objective robust optimization method for SRM drive systems to improve their robustness against manufacturing tolerances and imperfections in mass production.
(6) To present application-oriented design optimization methods for SRM drive systems and present two examples for the applications.

1.5 Organization of the Book

Chapter 2 presents an overview of the fundamentals of design and optimization for SRMs and drive systems. The basic working principles, magnetic fields, and energy efficiency are introduced. Modeling methods such as FEA and magnetic equivalent circuits are presented. Basic control methods are investigated, including the current chopping control, angular position control, direct torque control, and torque function control.

Chapter 3 introduces the conventional design methods and proposes a novel magnetic parameter design method based on the nonlinear characteristics of SRMs. Design examples are investigated to verify the proposed method and show its benefits.

Chapter 4 provides an overview of the control methods for SRMs, including conventional control, indirect control, direct control, and model predictive control. An example of improved direct torque control is investigated and presented.

Chapter 5 presents the optimization techniques for electrical machines from various aspects, including the design of experiments approach, sensitivity analysis, surrogate models, and optimization algorithms. Examples are presented in these sections as well.

Chapters 6 and 7 present the system-level deterministic and robust optimization methods for SRM drive systems. The optimization targets are multiple objectives, such as torque, loss, and torque ripple. Case studies are conducted to demonstrate the implementation and effectiveness of the proposed optimization methods.

Chapter 8 develops an application-oriented optimization method for SRMs for the application scenarios. Two kinds of applications (HEVs and EVs) are investigated.

Chapter 9 presents future trends for the design, optimization, and application of SRMs.

References

1. Papadis E, Tsatsaronis G. Challenges in the decarbonization of the energy sector. Energy. 2020;205. https://doi.org/10.1016/j.energy.2020.118025
2. Sorrell S. Reducing energy demand: a review of issues, challenges and approaches. Renew Sust Energ Rev. 2015;47:74–82. https://doi.org/10.1016/j.rser.2015.03.002.
3. Global EV Data Explorer. 2023. https://www.iea.org/data-and-statistics/data-tools/global-ev-data-explorer. Accessed.
4. (NOAA) NOaAA: Trends in Atmospheric Carbon Dioxide. https://www.esrl.noaa.gov/gmd/ccgg/trends/. Accessed.
5. Global Climate Change. https://data.giss.nasa.gov. Accessed.
6. Agency IE: Global Energy Review. 2021. https://www.iea.org/reports/global-energy-review-2021/electricity. Accessed.
7. Energy USDo. https://www.energy.gov/. Accessed.
8. EV-Volumes: Global Plug-In Vehicle Sales. https://www.ev-volumes.com/. Accessed.
9. Sanguesa JA, Torres-Sanz V, Garrido P, Martinez FJ, Marquez-Barja JM. A review on electric vehicles: technologies and challenges. Smart Cities. 2021;4(1):372–404.
10. Diao K, Sun X, Lei G, Bramerdorfer G, Guo Y, Zhu J. System-level robust design optimization of a switched reluctance motor drive system considering multiple driving cycles. IEEE Trans Energy Convers. 2021;36(1):348–57. https://doi.org/10.1109/tec.2020.3009408.
11. Sun X, Cao J, Lei G, Guo Y, Zhu J. A composite sliding mode control for SPMSM drives based on a new hybrid reaching law with disturbance compensation. IEEE Trans Transp Electrif. 2021;7(3):1427–36. https://doi.org/10.1109/TTE.2021.3052986.
12. Boldea I, Tutelea LN, Parsa L, Dorrell D. Automotive electric propulsion systems with reduced or no permanent magnets: an overview. IEEE Trans Ind Electron. 2014;61(10):5696–711. https://doi.org/10.1109/tie.2014.2301754.
13. Survey USG. https://www.usgs.gov/centers/nmic. Accessed.
14. Han S, Diao K, Sun X. Overview of multi-phase switched reluctance motor drives for electric vehicles. Adv Mech Eng. 2021;13(9). https://doi.org/10.1177/16878140211045195
15. Diao K, Sun X, Bramerdorfer G, Cai Y, Lei G, Chen L. Design optimization of switched reluctance machines for performance and reliability enhancements: a review. Renew Sust Energ Rev. 2022;168.
16. Diao K, Bramerdorfer G, Sun X, Yang Z, Han S. Multiobjective design optimization of a novel switched reluctance motor with unequal alternating stator yoke segments. IEEE Trans Transp Electrif. 2023;9(1):512–21. https://doi.org/10.1109/TTE.2022.3174337.
17. Feng L, Sun X, Bramerdorfer G, Zhu Z, Cai Y, Diao K, et al. A review on control techniques of switched reluctance motors for performance improvement. Renew Sust Energ Rev. 2024;199:114454. https://doi.org/10.1016/j.rser.2024.114454.
18. Borg Bartolo J, Degano M, Espina J, Gerada C. Design and initial testing of a high-speed 45-kW switched reluctance drive for aerospace application. IEEE Trans Ind Electron. 2017;64(2):988–97. https://doi.org/10.1109/tie.2016.2618342.
19. Feng L, Sun X, Guo D, Yao M, Diao K. Advanced torque sharing function strategy with sliding mode control for switched reluctance motors. IEEE Trans Transp Electrif. 2023;1–1. https://doi.org/10.1109/TTE.2023.3287159
20. Feng L, Sun X, Yang Z, Diao K. Optimal torque sharing function control for switched reluctance motors based on active disturbance rejection controller. IEEE/ASME Trans Mechatron. 2023;1–9. https://doi.org/10.1109/TMECH.2023.3240986

21. Xiao D, Ye J, Fang G, Xia Z, Wang X, Emadi A. A regional phase-locked loop-based low-speed position-sensorless control scheme for general-purpose switched reluctance motor drives. IEEE Trans Power Electron. 2022;37(5):5859–73. https://doi.org/10.1109/tpel.2021.3130273.
22. Gan C, Chen Y, Sun Q, Si J, Wu J, Hu Y. A position sensorless torque control strategy for switched reluctance machines with fewer current sensors. IEEE/ASME Trans Mechatron. 2021;26(2):1118–28. https://doi.org/10.1109/tmech.2020.3017864.
23. Oksuztepe E. In-wheel switched reluctance motor design for electric vehicles by using a Pareto-based multiobjective differential evolution algorithm. IEEE Trans Veh Technol. 2017;66(6):4706–15. https://doi.org/10.1109/tvt.2016.2618119.
24. Sun X, Tang X, Tian X, Lei G, Guo Y, Zhu J. Sensorless control with fault-tolerant ability for switched reluctance motors. IEEE Trans Energy Convers. 2022;37(2):1272–81. https://doi.org/10.1109/TEC.2021.3119928.
25. Sun X, Tang X, Tian X, Wu J, Zhu J. Position sensorless control of switched reluctance motor drives based on a new sliding mode observer using Fourier flux linkage model. IEEE Trans Energy Convers. 2022;37(2):978–88. https://doi.org/10.1109/TEC.2021.3125494.
26. Feng L, Sun X, Tian X, Diao K. Direct torque control with variable flux for an SRM based on hybrid optimization algorithm. IEEE Trans Power Electron. 2022;37(6):6688–97. https://doi.org/10.1109/TPEL.2022.3145873.
27. Takano Y, Chiba A, Ogasawara S, Takeno M, Hoshi N, Imakawa T, et al. Torque density and efficiency improvements of a switched reluctance motor without rare earth material for hybrid vehicles. In: 2010 IEEE energy conversion congress and exposition. 2010. p. 2653–9.
28. Takano Y, Takeno M, Hoshi N, Chiba A, Takemoto M, Ogasawara S, et al. Design and analysis of a switched reluctance motor for next generation hybrid vehicle without PM materials. In: The 2010 international power electronics conference—ECCE ASIA. 2010. p. 1801–6.
29. Yuichi T, Tomohiro M, Motoki T, Akira C, Nobukazu H, Masatsugu T, et al. Operating area of a switched reluctance motor with continuous current operation. IEEE PES general meeting. 2010. p. 1–4.
30. Chiba A, Takano Y, Takeno M, Imakawa T, Hoshi N, Takemoto M, et al. Torque density and efficiency improvements of a switched reluctance motor without rare-earth material for hybrid vehicles. IEEE Trans Ind Appl. 2011;47(3):1240–6. https://doi.org/10.1109/tia.2011.2125770.
31. Takeno M, Ogasawara S, Chiba A, Takemoto M, Hoshi N. Power and efficiency measurements and design improvement of a 50 kW switched reluctance motor for hybrid electric vehicles. In: 2011 IEEE energy conversion congress and exposition. 2011. p. 1495–501.
32. Imakawa T, Chimata K, Hoshi N, Chiba A, Takemoto M, Ogasawara S. Characteristic measurements of switched reluctance motor on prototype electric vehicle. In: 2012 IEEE international electric vehicle conference. 2012. p. 1–8.
33. Kiyota K, Kakishima T, Chiba A. Comparison of test result and design stage prediction of switched reluctance motor competitive with 60-kW rare-earth PM motor. IEEE Trans Ind Electron. 2014;61(10):5712–21. https://doi.org/10.1109/tie.2014.2304705.
34. Takase K, Goto H, Ichinokura O. Design and experiment of double stator structure axial-flux SR motor for electric city commuters. In: 2016 19th International conference on electrical machines and systems (ICEMS). 2016. p. 1–6.
35. Li SF, Zhang S, Habetler TG, Harley RG. Modeling, design optimization, and applications of switched reluctance machines-a review. IEEE Trans Ind Appl. 2019;55(3):2660–81. https://doi.org/10.1109/Tia.2019.2897965.
36. Sun X, Shi Z, Cai Y, Lei G, Guo Y, Zhu J. Driving-cycle-oriented design optimization of a permanent magnet hub motor drive system for a four-wheel-drive electric vehicle. IEEE Trans Transp Electrif. 2020;6(3):1115–25. https://doi.org/10.1109/TTE.2020.3009396.
37. Xue XD, Cheng KWE, Ng TW, Cheung NC. Multi-objective optimization design of in-wheel switched reluctance motors in electric vehicles. IEEE Trans Ind Electron. 2010;57(9):2980–7. https://doi.org/10.1109/TIE.2010.2051390.
38. Sun X, Diao K, Yang Z. Performance improvement of a switched reluctance machine with segmental rotors for hybrid electric vehicles. Comput Electr Eng. 2019;77:244–59. https://doi.org/10.1016/j.compeleceng.2019.06.002.

39. Lei G, Bramerdorfer G, Ma B, Guo Y, Zhu J. Robust design optimization of electrical machines: multi-objective approach. IEEE Trans Energy Convers. 2021;36(1):390–401. https://doi.org/10.1109/TEC.2020.3003050.
40. Ma B, Lei G, Zhu J, Guo Y, Liu C. Application-oriented robust design optimization method for batch production of permanent-magnet motors. IEEE Trans Ind Electron. 2018;65(2):1728–39. https://doi.org/10.1109/TIE.2017.2748046.
41. Sedrine EB, Ojeda J, Gabsi M, Slama-Belkhodja I. Fault-tolerant control using the GA optimization considering the reluctance torque of a five-phase flux switching machine. IEEE Trans Energy Convers. 2015;30(3):927–38. https://doi.org/10.1109/TEC.2015.2402234.
42. Mohammadi AS, Trovão JPF, Antunes CH. Component-level optimization of hybrid excitation synchronous machines for a specified hybridization ratio using NSGA-II. IEEE Trans Energy Convers. 2020;35(3):1596–605. https://doi.org/10.1109/TEC.2020.2990283.
43. Sun X, Shi Z, Lei G, Guo Y, Zhu J. Multi-objective design optimization of an IPMSM based on multilevel strategy. IEEE Trans Ind Electron. 2021;68(1):139–48. https://doi.org/10.1109/TIE.2020.2965463.
44. Guo Z, Liang Y, Bian X, Wang D. Multiobjective optimization for arrangement of the asymmetric-paths winding based on improved discrete particle swarm approach. IEEE Trans Energy Convers. 2018;33(3):1571–8.
45. Dong F, Zhao J, Song J, Feng Y, He Z. Optimal design of permanent magnet linear synchronous motors at multispeed based on particle swarm optimization combined with SN ratio method. IEEE Trans Energy Convers. 2018;33(4):1943–54.
46. Zhao W, Yao T, Xu L, Chen X, Song X. Multi-objective optimization design of a modular linear permanent-magnet Vernier machine by combined approximation models and differential evolution. IEEE Trans Ind Electron. 2020.
47. Rallabandi V, Wu J, Zhou P, Dorrell DG, Ionel DM. Optimal design of a switched reluctance motor with magnetically disconnected rotor modules using a design of experiments differential evolution FEA-based method. IEEE Trans Magn. 2018;54(11):1–5. https://doi.org/10.1109/TMAG.2018.2850744.
48. Diao K, Sun X, Lei G, Guo Y, Zhu J. Multiobjective system level optimization method for switched reluctance motor drive systems using finite-element model. IEEE Trans Ind Electron. 2020;67(12):10055–64. https://doi.org/10.1109/TIE.2019.2962483.
49. Ma C, Qu L. Multiobjective optimization of switched reluctance motors based on design of experiments and particle swarm optimization. IEEE Trans Energy Convers. 2015;30(3):1144–53. https://doi.org/10.1109/TEC.2015.2411677.
50. Zhang J, Wang H, Chen L, Tan C, Wang Y. Multi-objective optimal design of bearingless switched reluctance motor based on multi-objective genetic particle swarm optimizer. IEEE Trans Magn. 2018;54(1):1–13. https://doi.org/10.1109/TMAG.2017.2751546.
51. Du G, Huang N, Zhao Y, Lei G, Zhu J. Comprehensive sensitivity analysis and multiphysics optimization of the rotor for a high speed permanent magnet machine. IEEE Trans Energy Convers. 2021;36(1):358–67. https://doi.org/10.1109/TEC.2020.3005568.
52. Nabeta SI, Chabu IE, Lebensztajn L, Correa DAP, Silva WMd, Hameyer K. Mitigation of the torque ripple of a switched reluctance motor through a multiobjective optimization. IEEE Trans Magn. 2008;44(6):1018–21. https://doi.org/10.1109/TMAG.2007.915137
53. Xia B, Ren Z, Choi K, Koh C. A novel subregion-based multidimensional optimization of electromagnetic devices assisted by kriging surrogate model. IEEE Trans Magn. 2017;53(6):1–4. https://doi.org/10.1109/TMAG.2017.2655099.
54. Diao K, Sun X, Lei G, Guo Y, Zhu J. Multimode optimization of switched reluctance machines in hybrid electric vehicles. IEEE Trans Energy Convers. 2021;36(3):2217–26. https://doi.org/10.1109/tec.2020.3046721.
55. Lei G, Liu C, Zhu J, Guo Y. Techniques for multilevel design optimization of permanent magnet motors. IEEE Trans Energy Convers. 2015;30(4):1574–84. https://doi.org/10.1109/TEC.2015.2444434.
56. Li J, Li Y, Wang Y. Fuzzy inference NSGA-III algorithm-based multi-objective optimization for switched reluctance generator. IEEE Trans Energy Convers. 2021;36(4):3578–81. https://doi.org/10.1109/TEC.2021.3099961.

57. Bramerdorfer G, Tapia JA, Pyrhönen JJ, Cavagnino A. Modern electrical machine design optimization: techniques, trends, and best practices. IEEE Trans Ind Electron. 2018;65(10):7672–84. https://doi.org/10.1109/TIE.2018.2801805.
58. Diao K, Sun X, Lei G, Bramerdorfer G, Guo Y, Zhu J. Robust design optimization of switched reluctance motor drive systems based on system-level sequential Taguchi method. IEEE Trans Energy Convers. 2021;36(4):3199–207. https://doi.org/10.1109/tec.2021.3085668.
59. Ziyan R, Zhang D, Koh CS. Multi-objective worst-case scenario robust optimal design of switched reluctance motor incorporated with FEM and kriging. In: 2013 International conference on electrical machines and systems (ICEMS). 2013. p. 716–9.
60. Lei G, Wang T, Zhu J, Guo Y, Wang S. System-level design optimization method for electrical drive systems—robust approach. IEEE Trans Ind Electron. 2015;62(8):4702–13. https://doi.org/10.1109/TIE.2015.2404305.

Chapter 2
Fundamentals of Switched Reluctance Motors

Abstract This chapter briefly summarizes the design fundamentals for switched reluctance motors (SRMs), including the working principles, modeling methods, and novel topologies. The modeling methods are employed to analyze SRMs' performances based on their nonlinear characteristics induced by the working principle. These methods will be used to optimize SRMs and drive systems in the following chapters.

Keywords Design method · Finite element analysis · Magnetic equivalent circuit · Modeling method · Switched reluctance motors

2.1 Introduction

The operational principles of switched reluctance motors (SRMs) are based on the variable reluctance effect, by which the magnetic flux seeks the path of least reluctance. The SRM has a stator with concentrated phase windings and a rotor with salient poles. When the stator phase windings are energized sequentially, the stator magnetic field will attract the rotor poles to minimize the airgap reluctance between the stator and rotor, which causes the rotor to rotate. Various modeling techniques have been used to analyze and optimize the performance of SRMs, including analytical, numerical, and experimental methods. The control strategies for SRMs have evolved from conventional approaches, such as current chopping control and angular position control, to advanced techniques, such as fuzzy logic and artificial intelligence-based methods [1]. The development of novel topologies, such as multi-phase and hybrid SRMs, has also expanded the range of design options for SRMs. Moreover, according to different standards, the optimization has multiple classification methods [2, 3].

This chapter briefly introduces the working principle, analysis methods, and various topologies for SRMs. The following sections are organized as follows. Section 2.2 provides an overview of the working principles of SRMs. Section 2.3 covers the modeling methods used for SRMs, including analytical, numerical, and

X. Sun et al., *Multi-objective Design Optimization of Switched Reluctance Motor Drive Systems*, https://doi.org/10.1007/978-981-96-0672-6_2

experimental approaches. Section 2.4 describes novel SRM topologies for SRMs, followed by the summary in Sect. 2.5.

2.2 Working Principle of Switched Reluctance Motors

2.2.1 Operational Principle

The operation of an SRM is based on the concept that a magnetic circuit's reluctance depends on the length and cross-sectional area of the path through which the magnetic flux flows [4]. When a stator phase winding is energized, the magnetic flux flows through the stator pole, the air gap, and the rotor pole. Suppose the rotor pole is misaligned with the stator pole (Fig. 2.1a). The stator magnetic field will attract the rotor pole to align the stator and rotor poles to minimize the airgap magnetic reluctance between them (Fig. 2.1b). The rotor will rotate continuously when the stator poles are sequentially energized in a predetermined sequence. The speed of the motor is controlled by adjusting the frequency and duration of the current pulses applied to the stator windings.

The design of an SRM refers to determining optimal stator and rotor shapes and dimensions to ensure efficient and effective motor operation. The number of stator and rotor poles and the alignment of the poles are critical factors in determining motor performance. The stator and rotor pole shapes affect the magnetic circuit and magnetic flux distribution, which impacts the motor torque and power output. The winding configuration and the switching pattern of the stator windings also affect the motor performance. The winding pattern and the stator and rotor pole shapes should be carefully designed to ensure a smooth variation of the airgap magnetic reluctance between the stator and rotor poles during their alignment to minimize the torque ripple. The switching pattern should also be optimized to ensure that the current pulses applied to the stator windings are timed correctly to achieve maximum torque and efficiency.

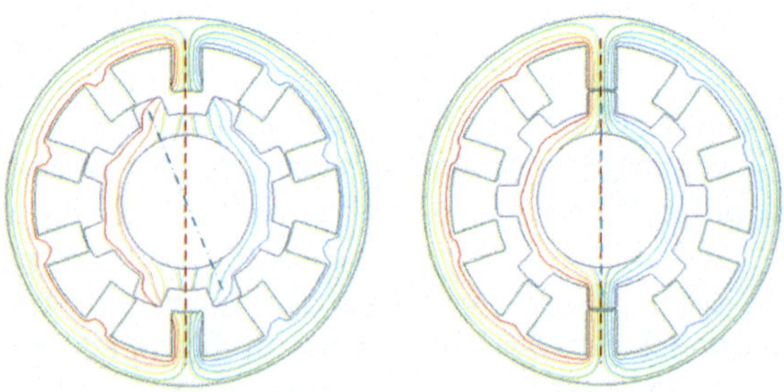
(a) (b)

Fig. 2.1 **a** Unaligned and **b** aligned positions of the SRM

2.2.2 Torque Generation

The voltage balance equation of an SRM phase winding can be expressed as

$$v = iR + \frac{\mathrm{d}\lambda}{\mathrm{d}t} \tag{2.1}$$

where v, i, R, and λ are the voltage, current, resistance, and flux linkage of the stator phase winding.

The winding flux linkage can be related to the phase winding excitation current as follows:

$$\lambda = L(\theta)i \tag{2.2}$$

where $L(\theta)$ is the phase winding self-inductance when the rotor is positioned at angular position θ in rad (mechanical).

Substituting (2.2) into (2.1), one obtains

$$v = iR + L(\theta)\frac{\mathrm{d}i}{\mathrm{d}t} + i\frac{\mathrm{d}L(\theta)}{\mathrm{d}\theta}\omega_r \tag{2.3}$$

where

$$\omega_r = \frac{\mathrm{d}\theta}{\mathrm{d}t} \tag{2.4}$$

is the rotor speed in rad/s (mechanical).

Multiplying both sides of (2.4) by phase current, i, one obtains the power balance equation of an SRM as follows:

$$iv = i^2R + iL(\theta)\frac{\mathrm{d}i}{\mathrm{d}t} + i^2\frac{\mathrm{d}L(\theta)}{\mathrm{d}\theta}\omega_r. \tag{2.5}$$

Assume that the SRM works in the magnetic linear region. The second term on the right-hand side term of (2.5) can be expanded as:

$$iL(\theta)\frac{\mathrm{d}i}{\mathrm{d}t} = \frac{\mathrm{d}}{\mathrm{d}t}\left(\frac{1}{2}L(\theta)i^2\right) - \frac{1}{2}i^2\frac{\mathrm{d}L(\theta)}{\mathrm{d}\theta}\omega_r. \tag{2.6}$$

Substituting (2.6) into (2.5), one obtains:

$$iv = i^2R + \frac{\mathrm{d}}{\mathrm{d}t}\left(\frac{1}{2}L(\theta)i^2\right) + \frac{1}{2}i^2\frac{\mathrm{d}L(\theta)}{\mathrm{d}\theta}\omega_r. \tag{2.7}$$

Equation (2.7) shows that the input electric power fed into an SRM is converted into three parts: electrical power, magnetic power, and mechanical power. The electrical power is the copper loss dissipated in the phase winding. Magnetic power is the variation rate of magnetic energy stored in the magnetic core. The mechanical power is translated into torque production. Thus, the torque in the linear region can be expressed as:

$$T = \frac{1}{2} i^2 \frac{\mathrm{d}L(\theta)}{\mathrm{d}\theta}. \tag{2.8}$$

It should be noted that the general torque equation for SRM is derived from magnetic co-energy, which is applicable for both magnetically linear and saturated regions:

$$T = \left. \frac{\partial W_c}{\partial \theta} \right|_{i=const}. \tag{2.9}$$

2.3 Modeling Methods of Switched Reluctance Motors

Accurate electromagnetic modeling provides the foundation for the subsequent design and optimization processes. A number of studies considered the electromagnetic performances of different types of machines. These approaches can be classified into two categories: numerical and analytical methods.

2.3.1 Numerical Methods

Numerical methods are commonly used to offer a trustworthy electromagnetic solution for various electrical machines. There are two main types of numerical methods: the finite element analysis (FEA) and the boundary element method (BEM).

(a) FEA

FEA is the most promising and popular numerical method for electrical machines. In FEA, the analysis domain is divided into a mesh of smaller, simpler pieces called finite elements. FEA assumes a suitable approximate solution for each element, and the solution of this domain, which satisfies the conditions like the equilibrium condition of the structure, is deduced. Dirichlet, Neumann, and periodic boundary conditions are applied for solving FEA problems. The definition and assignment of finite elements are flexible. Thus, FEA suits complex machine topologies and can provide accurate performance solutions, like magnetic field distribution. The computational burden is the main limit of FEA since it needs to mesh the entire solution domain. With the increased computational power, this problem has been gradually improved, making FEA the most commonly used tool. Two-dimensional (2-D) FEA instead of three-dimensional (3-D) FEA is commonly used and has become a mature technology applied to design SRMs. The major deficiency of 2-D FEA is that it is generally impossible to model the variations of material properties along the axial direction and skewing and end winding effects [5, 6]. Consequentially, for complex topologies, especially machines with axial flux paths or skewed components, 3-D FEA provides more accurate results [7].

(b) BEM

BEM is an alternative numerical method to obtain a precise magnetic solution, which also has been applied in the design optimization of SRMs [8, 9]. Similar to FEA, several equations are assembled into a matrix form and solved to obtain the solution. In contrast to FEA, BEM forms the boundary value problems as integral equations with boundary conditions rather than differential equations throughout the whole design domain. The main drawbacks of BEM are that the system matrix lacks symmetry and its ability to deal with nonlinear problems is poor due to the coarse treatment of anisotropic material properties. Accordingly, it is usually combined with FEA or magnetic equivalent circuit (MEC) [10–12].

2.3.2 *Analytical Methods*

Compared with numerical methods, analytical techniques can significantly reduce the computational effort, often at the expense of accuracy. Moreover, numerical methods cannot give a direct insight into the influence of geometrical parameters on the machine's performance. Thus, this kind of method could not explain the natural characteristics well. In contrast, analytical techniques facilitate studying those circumstances. The main challenge of the analytical method is that it cannot approximate a complex geometry well by taking an accurate evaluation of core saturation and end-winding inductance into account. The analytical methods can be classified into Maxwell-equations-based, curve-fitting, and MEC methods.

(a) Maxwell-equations-based methods

Maxwell-equations-based methods provide magnetic field analysis by solving the partial differential equations of magnetic potentials based on Maxwell's equations. In [13], an analytical model was developed for the unaligned inductance of the SRM. The contributions of stator and rotor slots have been accounted for. However, the stator and rotor are reshaped into rectangles for simplification, and they cannot be applied to other positions where the iron core is saturated. An enhanced model was presented in [14] to calculate the phase inductance over the full rotor position ranges. Conformal mapping is employed to calculate the permanence of the air region, and the saturation effects are considered by combining them with the MEC model. It has been proved that the slotting effects can be reduced using harmonic modeling and the Schwarz-Christoffel mapping method [15].

(b) Curve-fitting methods

Curve-fitting methods approximate the models like the flux linkage or static torque with respect to rotor position and phase current. It estimates the profiles of SRM's characteristics based on a limited set of data from simulation or experiments rather than a complete set of data, which can somewhat reduce the modeling cost. There are four main types of curve-fitting methods, i.e., a Fourier series-based approach, lookup-based versions, linear/nonlinear curve-fitting formulas, and the application of the artificial intelligence method, which all have been successfully applied in SRMs [16].

The Fourier series-based method has been extensively used to characterize the nonlinearity of flux linkage, inductance, or torque of SRMs. The accuracy of the model typically increases with the number of considered orders. In [17], the flux-linkage models were represented by the second- and fourth-order Fourier series, respectively, as presented in Fig. 2.2. It can be found that in the second-order model, the flux linkage at the aligned position is lower than the adjacent positions, which obey the consensus. Furthermore, the initial data for Fig. 2.2 is based on the experimental results of torque-balanced positions where the static torque is zero. Thus, the cost can be greatly reduced compared to the entire measurement range.

(c) MEC

MEC is an efficient tool that provides a good compromise between computational burden and modeling accuracy. It is analogous to an electric circuit with a reluctance network representing the electrical machine. The level of physical geometry discretization mainly characterizes the accuracy of MEC. The reluctance of each region depends on the geometry and material permeability. The reluctance value regarding the airgap region continuously varies with the rotor position. Generally, assume that a flux tube has two end surfaces (S_1 and S_2), as shown in Fig. 2.3. The reluctance of this flux tube R_{m} can be calculated by

$$R_{\mathrm{m}} = \int_0^l \frac{\mathrm{d}x}{\mu(x)S(x)} = \frac{V_M}{\Phi} \tag{2.10}$$

where l is the length of the flux tube, $\mu(x)$ is the nonlinear permeability, $S(x)$ is the variable cross-sectional area, V_M is the magnetomotive force, and Φ is the flux linkage.

Figure 2.4 presents an MEC model for an 8/6 SRM, investigated in [18]. There are three main types of reluctance, i.e., linear reluctance with constant permeability, nonlinear reluctance decided by the material property, and variable reluctance corresponding to the rotor position. Both the stator and rotor are divided into two types of reluctance. For instance, as shown in Fig. 2.4, R_{sy} and R_{st} represent the reluctances

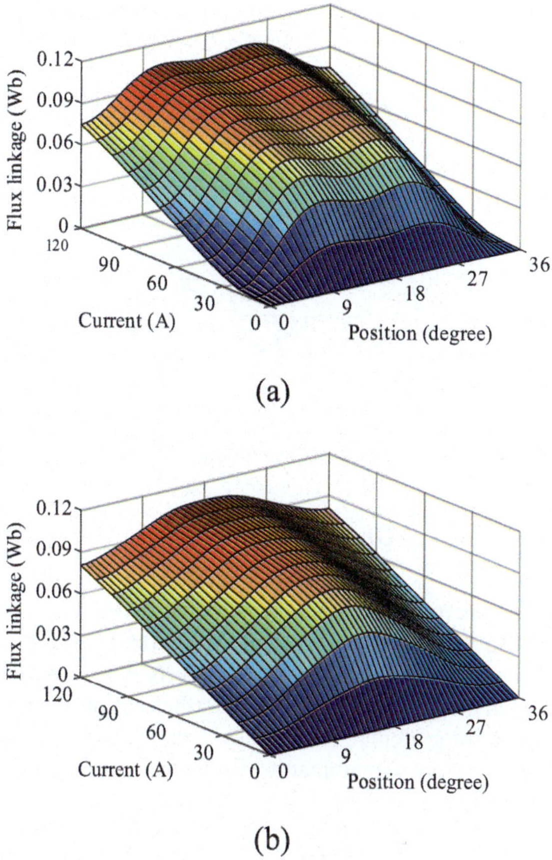

Fig. 2.2 Flux linkage modeling by **a** second-order and **b** fourth-order Fourier series [17]

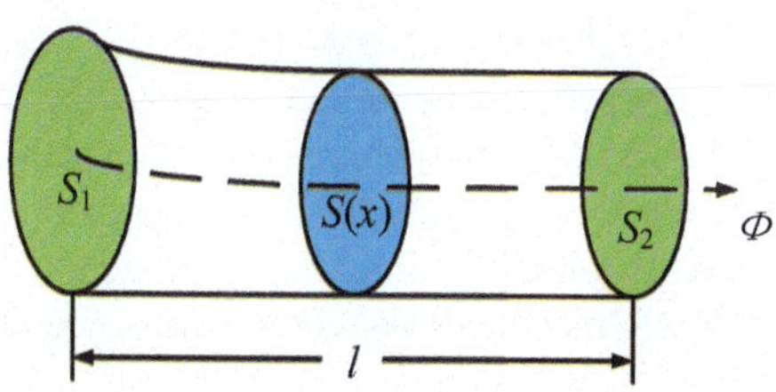

Fig. 2.3 Definition of the flux tube

of the stator yoke and teeth, and R_{ry} and R_{rt} represent the reluctances of the rotor yoke and teeth, respectively. The flux passes through the stator and rotor iron cores would be nonuniform with the rotation of the rotor. Therefore, more discretization of the stator and rotor region is beneficial to improve the modeling accuracy.

The aforementioned magnetic modeling methods for SRMs are summarized and compared in Table 2.1 for better visualization and understanding.

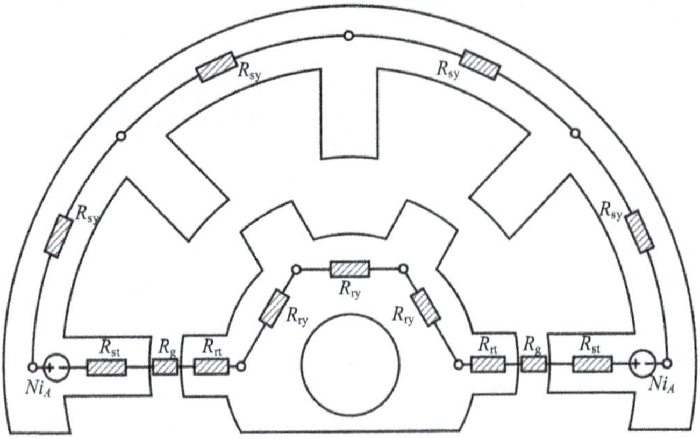

Fig. 2.4 An MEC model of an 8/6 SRM [18]

2.4 Novel Design Topologies

Over the past decades, conventional SRMs have been successfully designed and tested in industry applications. Moreover, various novel topologies are derived from conventional structures for further performance and reliability improvements or some specific applications. The novel structures of SRMs can be divided into segmented stator/rotor, double stator/rotor, hybrid excitation, skewed teeth, axial flux, linear structures, or a combination of two or several of these types.

2.4.1 Segmented SRMs

The flux path can be further shortened by incorporating the segmented parts into the SRMs. Besides, the electromagnetic isolation caused by the segmented structures can enhance the reliability and fault tolerance of the motors. The segmented SRMs can be generally classified into the segmented-stator SRMs (SSSRMs) and the segmented-rotor SRMs (SRSRMs). It could also be the combination of segmented stators and rotors.

There are various topologies of SSSRMs, and they can be divided into SSSRMs with C-core, E-core, and multi-tooth stators according to their tooth number of segmented stators [34–36]. For instance, as reported in [37] and [38], a novel two-phase E-core SSSRM was proposed and investigated. The stator was E-shaped with wide and narrow teeth, as presented in Fig. 2.5a. Compared with the conventional two-phase SRMs, the material for producing the stator core can be saved, and the iron loss can be reduced. An improved 9/12 structure SRM was proposed in [39] to produce higher output torque. The impact of the number of segmented-stator teeth

Table 2.1 Summary of magnetic modeling methods for SRMs

Classification	Modeling approach	Reference	Advantages	Disadvantages
Numerical methods	FEA	[5–7, 19–23]	High accuracy; suitable for complex structures, accurate field distributions, and good consideration of 3-D effects, such as skewing and winding effects	High computational cost; complex modeling process for 3-D geometry
	BEM	[8, 9]	High accuracy; moderate computation cost	Dense system matrix; poor ability to deal with saturation or nonlinear problems
Analytical methods	Maxwell-equations-based method	[13, 14, 24–26]	Low computation cost; clear magnetic characteristics	High complexity; poor ability to deal with saturation or nonlinear problems
	Curve-fitting method	[17, 27–30]	Low computation cost based on a limited number of data	Empirical and heuristic; requiring preset data; new data is required if the topology is changed
	MEC	[12, 18, 31–33]	Acceptable accuracy; low computation cost	Empirical assumptions of fringing/leakage reluctances; requirement to define flux paths in advance

on the performance was investigated in [40], and a novel SSSRM with a multi-tooth stator was proposed to maximize the converted energy, as shown in Fig. 2.5b.

In the literature [41], Mecrow et al. presented a three-phase SRSRM with full-pitch windings, as shown in Fig. 2.6a. The rotor comprises several discrete fan-shaped core blocks which are embedded in the non-magnetic sleeve. The defect of

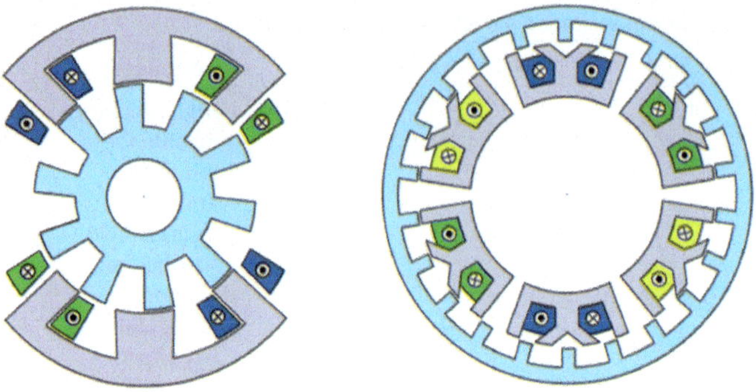

(a) (b)

Fig. 2.5 Machine topologies of SSSRMs with **a** E-core [37, 38] and **b** multi-tooth stator [40]

this structure is that the windings span multiple stator teeth so that the windings of different phases overlap at the end of the stator, which increases the end effects and reduces the fault tolerance. Thus, it is not suitable for applications with small axial lengths. To solve this problem, an SRSRM with two types of stator poles was proposed in [42]. The windings of this topology were placed around a single tooth, which occupies less copper volume than the full-pitch SRSRM, as shown in Fig. 2.6b. The performances of the SRSRMs were evaluated and compared in [43]. It was concluded that both SRSRMs with fully pitched and single-tooth windings could deliver 40% higher output torque than conventional SRMs at the same copper consumption. The SRSRMs with single-tooth windings would be more attractive due to their lower end-windings. Moreover, in [44], the SRSRMs with exciting and auxiliary stator poles were proposed, and the effects of different rotor pole numbers were investigated.

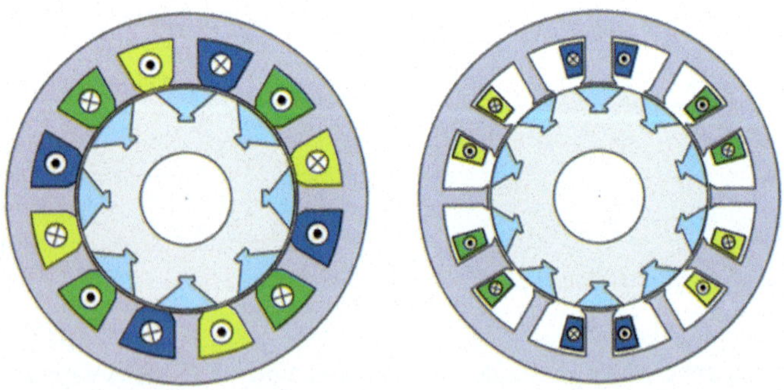

(a) (b)

Fig. 2.6 Machine topologies of **a** an SRSRM with full-pitch windings, and **b** an SRSRM with single-tooth windings [41]

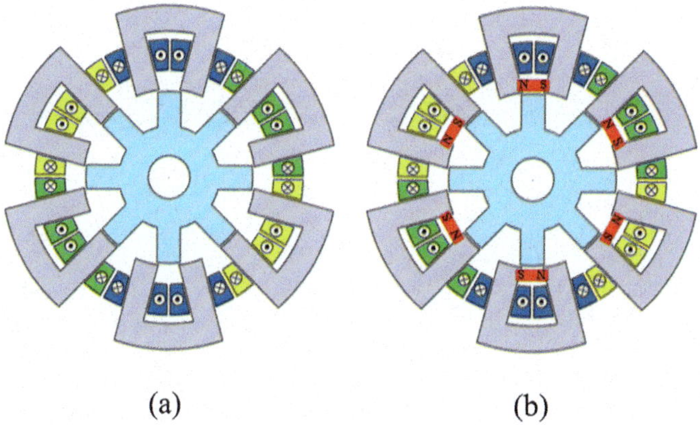

Fig. 2.7 Machine topologies of **a** a C-core SRM and **b** an HESRM [45]

2.4.2 Hybrid-Excitation SRMs

Compared with PM motors, conventional SRMs exhibit lower torque and power density. One solution to this issue is employing auxiliary flux sources to improve the flux density. Therefore, hybrid-excitation SRMs (HESRMs) are gaining increasing attention by adding PMs to improve torque production and power density.

In [45], a 12/8 HESRM was presented based on a C-core SSSRM, as shown in Fig. 2.7. It is found that the HESRM has higher torque than a conventional SRM of the same size. Furthermore, an investigation has been performed in [46] to present the influence of this topology with different pole numbers. Two HESRMs with 12/10 and 12/8 stator/rotor configurations were manufactured and comprehensively compared. The problem with such HESRM (where the PMs are placed between the stator pole tips) is that the winding area is narrowed, which will influence the maximum output torque. Moreover, the heat dissipation performance gets worse. An alternative way is to arrange the PMs inside the stator teeth. However, it will cause insufficient mechanical strength [47]. Another innovative method of embedding PMs has been proposed in recent work [48]. A novel multi-tooth HESRM was presented, where the PMs are placed between the end teeth of the adjacent modules. It can well handle the problems of the two other solutions where PMs are placed between pole tips or inside stator teeth.

2.4.3 Double-Stator/Rotor SRMs

Conventional SRMs suffer from the high vibration caused by the radial force. The double-stator SRM (DSSRM) and double-rotor SRM (DRSRM) presented in Fig. 2.8 are effective solutions since the electromagnetic force from the opposite stator poles

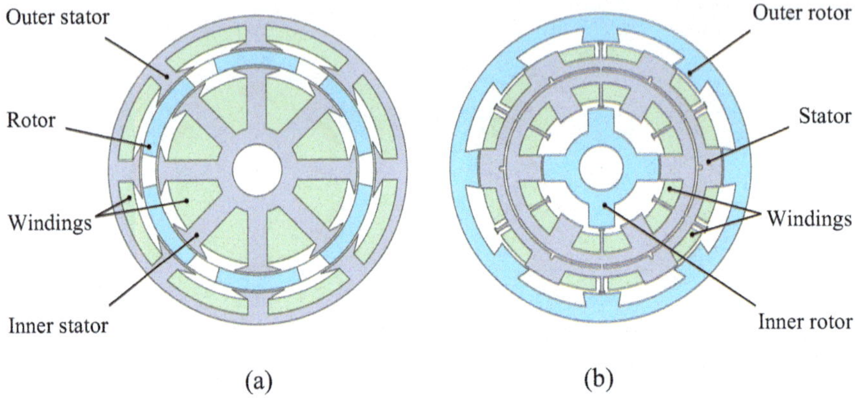

Fig. 2.8 Machine topologies of **a** a DSSRM [49] and **b** a DRSRM [59]

can be offset to some extent. Besides, both the DSSRM and DRSRM have been proposed as high-power density motors.

The DSSRM presented in Fig. 2.8a was first introduced in [49], and the configuration and the magnetic force analysis were presented. The vibration characteristics were reported in [50] and compared with a conventional SRM. It was found that the vibration could be significantly reduced. Besides, the thermal analysis of DSSRM was performed in [51]. The design strategy, including the calculation of dimensions and electromagnetic analysis of DSSRM, was proposed in [52]. In [53], a novel mechanical offset DSSRM with multi-tooth was proposed. The two stators can provide complementary torques when teeth of the outer and inner rotor are deliberately offset by conjugate angle, hence minimizing the torque ripple. Recent works regarding the DSSRM focus on modifying and optimizing the configurations for torque ripple mitigation [21, 54, 55].

As reported in [56], the DRSRM was first proposed along with the working principle. Since it could provide two independent mechanical ports, DRSRM is especially suitable for HEV transmission systems where dual electromechanical energy conversion is required. A segmented rotor was incorporated into the DRSRM to improve torque production [57]. Based on those results, a unique DRSRM with a shared back-iron was presented in [58] to reduce the overall size. Yang et al. [59] presented the detailed design procedure of the DRSRM, including the analytical calculations and FEA. A novel DRSRM, presented in Fig. 2.8b, was designed and optimized so that both rotors could be controlled independently. Moreover, a family of DRSRM configurations has been introduced and compared.

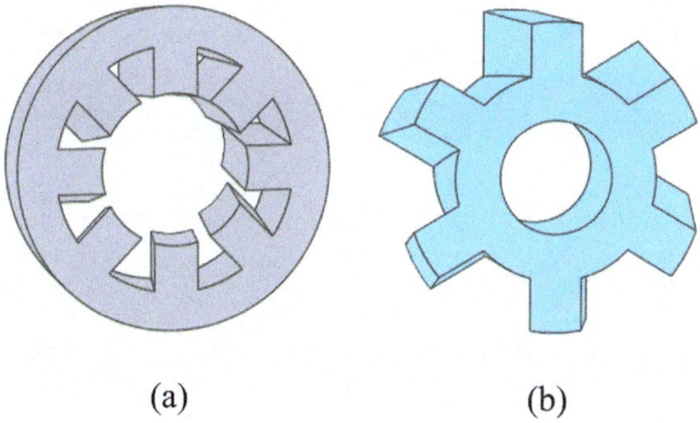

Fig. 2.9 Topologies of skewed-teeth SRMs, **a** stator and **b** rotor [60]

2.4.4 *Skewed-Teeth SRMs*

Figure 2.9 presents the stator and rotor topologies of skewed SRMs. Considering the skewed stator and rotor combination, the skewed-teeth SRMs (STSRMs) can be divided into three categories: stator-only, rotor-only, and stator-rotor combination skewing structures. These SRMs suffer from large vibrations due to the radial magnetic force generated between the double salient poles. The benefit of the skewed topology is that it can reduce this vibration by mitigating the radial force [60].

The effects of the three types of STSRMs have been investigated in [61]. It was concluded that skewing can reduce the vibration caused by the radial pulsating force, and skewing the stator is more effective than skewing the rotor for vibration suppression. Moreover, skewing the stator and rotor together is the most effective way to reduce the vibration. Furthermore, in [62], the torque production of skewed-teeth SRMs was investigated and compared, and it was found that the stator-only and rotor-only skewing structures have negative effects on output torque, while the stator-rotor combination skewing structure has less impact.

2.4.5 *Axial-Flux SRM*

The magnetic flux lines in the air gap of the axial-flux SRM (AFSRM) are parallel to the rotation axis of the machine. It has been recognized that AFSRM exhibits higher power density than radial-flux SRM (RFSRM) due to its larger air gap surface [63]. The air gap surface of the former is dependent on the diameter of the machine, whereas that of the latter is dependent on the axial length of the machine. Accordingly, AFSRM is preferred over RFSRM for applications like in-wheel motors with a high diameter and axial length ratio. The basic properties of AFSRM were presented in [64]. It has been found that the airgap length of AFSRM must be larger than that of RFSRM since the AFSRM has a large electromagnetic force in the axial direction, which generates rotor distortion. An AFSRM with C-core stators and segmented rotors was

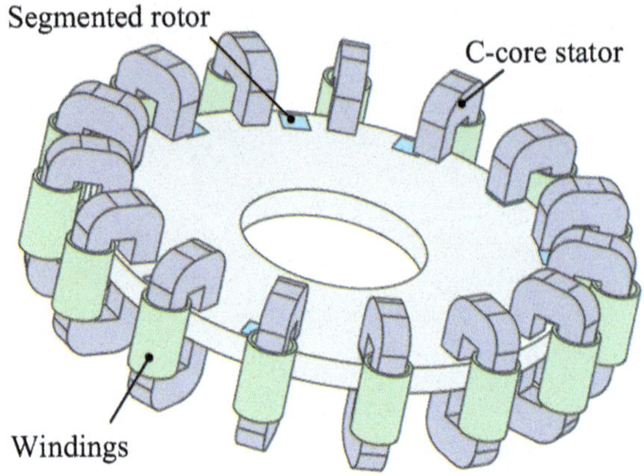

Fig. 2.10 Topology of an AFSRM [65]

proposed in [65], as shown in Fig. 2.10. It exhibits some major merits, such as higher power density, more flexible space for winding arrangement, maximized thermal dissipation, and improved fault tolerance capability.

2.4.6 Linear SRMs

In conventional linear motion systems, rotary machines are usually used to drive the connecting rod to convert the rotary motion into linear motion. The overall efficiency will be affected due to the energy loss in the multistage transmission mechanism. Thus, linear machines are superior to rotary machines in terms of stiffness, working life, and other performance indices for vertical actuation systems. Two features should be intensively observed for the application of linear motors. One is the payload ratio to the on-load capability, and the other is the feasibility of a long stator structure [66]. The linear machines could achieve a high respective ratio while simultaneously guaranteeing high efficiency. However, they both have windings in the stator, which are not feasible for long-distance propulsion due to the high cost.

Linear SRMs (LSRMs) are an attractive alternative and have been designed and applied to rail propulsion, wave energy generation, and elevators due to the absence of PMs, concentrated rather than distributed windings, and high robustness and fault tolerance. LSRMs can be classified into two types: longitudinal flux type or transverse flux type. In longitudinal flux LSRM, the plane of the flux is parallel to the direction of motion, while that of the transverse flux LSRM is perpendicular to the direction of motion. The design procedure of longitudinal flux LSRM by converting the specifications of LSRM into the equivalent rotary SRM that was proposed in [67]. Four longitudinal flux LSRM configurations were designed and compared in [68] for application in vertical elevators. In addition, the electromagnetic force characteristics of a transverse flux LSRM were investigated in [69]. The equations regarding the motors' force, efficiency, operating velocity, and required voltage have been presented in [70].

In recent years, the segmented stator/translator has been adopted in LSRMs. Wang et al. [66] presented the design, sensitivity analysis, and optimization of an LSRM with segmented rotors, as presented in Fig. 2.11a. It was concluded that the presented LSRM exhibits high force density, copper utilization ratio, and low cost. Furthermore, the unitized design methodology for this topology has been developed in [71]. Topologies for LSRM with segmented translator were introduced in [72], and an analytical modeling method has been developed for this kind of structure, as illustrated in Fig. 2.11b [25].

For better visualization, in Table 2.2, all the above novel topologies of SRMs have been summarized and compared. The features of each topology have been clearly illustrated, and their main advantages and disadvantages have been listed.

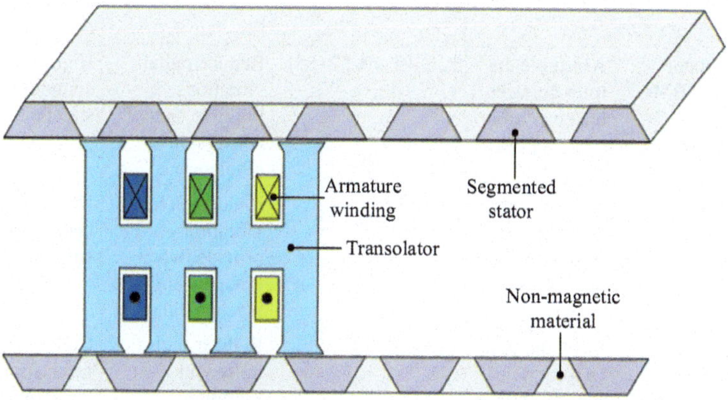

(a)

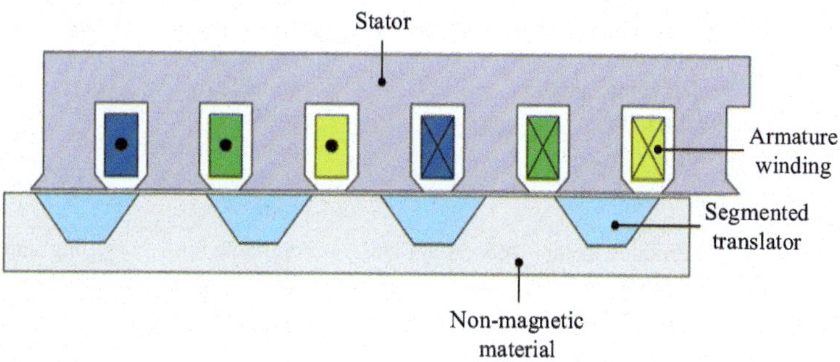

(b)

Fig. 2.11 Machine topologies of **a** a segmented-stator LSRM [66] and **b** a segmented-translator LSRM [25]

Table 2.2 Summary and comparison of various SRM topologies

Topology	Feature	Reference	Advantages	Disadvantages
Segmented-stator SRM (SSSRM)	Adopt the segmented stator	[34, 37–40, 73]	Reduce iron loss; improve output torque and fault-tolerance ability	High manufacturing and assembling complexity
Segmented-rotor SRM (SRSRM)	Adopt the segmented rotor	[6, 41–44, 74–76]	Improve output torque; reduce windage loss and torque ripple	High manufacturing complexity; potential mechanical weakness
Hybrid-excitation SRM (HESRM)	Add the PMs for flux enhancement	[45–48, 77, 78]	Improve output torque	Insufficient mechanical strength; less inherent robustness
Double-stator SRM (DSSRM)	Assemble the rotor between the outer and inner stators	[21, 49, 50, 52–55]	Reduce radial vibrations; improve energy conversion efficiency	High cost; complex structure
Double-rotor SRM (DRSRM)	Assemble the stator between the outer and inner rotors	[56–59]	Reduce radial vibration; provide two independent mechanical ports	High cost; complex structure
Skewed-teeth SRM (STSRM)	Skew the stator or rotor poles with the proper angle	[60–62, 79]	Reduce acoustic noise and vibration; controlled with conventional schemes	Complex fabrications with different stacks between two laminations
Axial-flux SRM (AFSRM)	The magnetic flux flows along the axial direction	[63–65, 80–82]	Produce high power density; suitable for applications with a high diameter to axial length ratio	High manufacturing cost
Linear SRM (LSRM)	Produce linear motion between stator and rotor	[66–68, 83–86]	Suitable for vertical actuation systems; high fault-tolerance ability	Long stator yoke causes difficulties in the manufacturing and assembling process

2.5 Summary

This chapter presented the foundations of SRMs. The operating principle and torque generation law were introduced. Modeling methods were categorized and discussed. Various novel design topologies, including segmented, hybrid-excitation, double-stator/rotor, skewed-teeth, axial-flux and linear SRMs, have been presented and compared to show their features.

References

1. Xiao D, Filho SR, Fang G, Ye J, Emadi A. Position-sensorless control of switched reluctance motor drives: a review. IEEE Trans Transp Electrif. 2022;8(1):1209–27. https://doi.org/10.1109/tte.2021.3110867.
2. Lei G, Bramerdorfer G, Ma B, Guo Y, Zhu J. Robust design optimization of electrical machines: multi-objective approach. IEEE Trans Energy Convers. 2021;36(1):390–401. https://doi.org/10.1109/TEC.2020.3003050.
3. Diao K, Sun X, Bramerdorfer G, Cai Y, Lei G, Chen L. Design optimization of switched reluctance machines for performance and reliability enhancements: a review. Renew Sust Energ Rev. 2022;168. https://doi.org/10.1016/j.rser.2022.112785
4. Krishnan R. Switched reluctance motor drives: modeling, simulation, analysis, design, and applications. CRC Press; 2017.
5. Yao W, Jin J, Krein PT. A highly efficient domain decomposition method applied to 3-D finite-element analysis of electromechanical and electric machine problems. IEEE Trans Energy Convers. 2012;27(4):1078–86. https://doi.org/10.1109/TEC.2012.2216528.
6. Diao K, Sun X, Lei G, Guo Y, Zhu J. Multiobjective system level optimization method for switched reluctance motor drive systems using finite-element model. IEEE Trans Ind Electron. 2020;67(12):10055–64. https://doi.org/10.1109/TIE.2019.2962483.
7. Lin J, Schofield N, Emadi A. External-rotor 6–10 switched reluctance motor for an electric bicycle. IEEE Trans Transp Electrification. 2015;1(4):348–56. https://doi.org/10.1109/TTE.2015.2502543.
8. Tang Y, Kline JA. Modeling and design optimization of switched reluctance machine by boundary element analysis and simulation. IEEE Trans Energy Convers. 1996;11(4):673–80. https://doi.org/10.1109/60.556360.
9. Fauchez MB. Magnetic analysis of a switched reluctance motor using a boundary element-finite element coupling method. IEEE Trans Magn. 1988;24(1):475–8. https://doi.org/10.1109/20.43960.
10. Araujo DM, Coulomb J, Chadebec O, Rondot L. A hybrid boundary element method-reluctance network method for open boundary 3-D nonlinear problems. IEEE Trans Magn. 2014;50(2):77–80. https://doi.org/10.1109/TMAG.2013.2281759.
11. O'Connell TC, Krein PT. A time-harmonic three-dimensional vector boundary element model for electromechanical devices. IEEE Trans Energy Convers. 2010;25(3):606–18. https://doi.org/10.1109/TEC.2010.2042811.
12. Sun X, Diao K, Lei G, Guo Y, Zhu J. Real-time HIL emulation for a segmented-rotor switched reluctance motor using a new magnetic equivalent circuit. IEEE Trans Power Electron. 2020;35(4):3841–9. https://doi.org/10.1109/TPEL.2019.2933664.
13. Radun A. Analytical calculation of the switched reluctance motor's unaligned inductance. IEEE Trans Magn. 1999;35(6):4473–81. https://doi.org/10.1109/20.809140.
14. Li S, Zhang S, Gong C, Habetler TG, Harley RG. An enhanced analytical calculation of the phase inductance of switched reluctance machines. IEEE Trans Ind Appl. 2019;55(2):1392–407. https://doi.org/10.1109/TIA.2018.2879303.

15. Connell TCO, Krein PT. A Schwarz–Christoffel-based analytical method for electric machine field analysis. IEEE Trans Energy Convers. 2009;24(3):565–77. https://doi.org/10.1109/TEC.2009.2025412.
16. Watthewaduge G, Sayed E, Emadi A, Bilgin B. Electromagnetic modeling techniques for switched reluctance machines: state-of-the-art review. IEEE Open J Ind Electron Soc. 2020;1:218–34. https://doi.org/10.1109/OJIES.2020.3016242.
17. Sun X, Diao K, Lei G, Guo Y, Zhu J. Direct torque control based on a fast modeling method for a segmented-rotor switched reluctance motor in HEV application. IEEE J Emerg Sel Top Power Electron. 2021;9(1):232–41. https://doi.org/10.1109/jestpe.2019.2950085.
18. Fleming FE, Edrington CS. Real-time emulation of switched reluctance machines via magnetic equivalent circuits. IEEE Trans Ind Electron. 2016;63(6):3366–76. https://doi.org/10.1109/TIE.2016.2521343.
19. Eit MA, Dular P, Bouillault F, Marchand C, Krebs G. Perturbation finite element method for efficient copper losses calculation in switched reluctance machines. IEEE Trans Magn. 2017;53(6):1–4. https://doi.org/10.1109/TMAG.2017.2655339.
20. Ma C, Qu L. Multiobjective optimization of switched reluctance motors based on design of experiments and particle swarm optimization. IEEE Trans Energy Convers. 2015;30(3):1144–53. https://doi.org/10.1109/tec.2015.2411677.
21. Wei W, Mengying L, Cosoroaba E, Fahimi B, Kiani M. Rotor shape investigation and optimization of double stator switched reluctance machine. IEEE Trans Magn. 2015;51(3):1–4. https://doi.org/10.1109/tmag.2014.2356573.
22. Anvari B, Toliyat HA, Fahimi B. Simultaneous optimization of geometry and firing angles for in-wheel switched reluctance motor drive. IEEE Trans Transp Electrification. 2018;4(1):322–9. https://doi.org/10.1109/TTE.2017.2766452.
23. Sun X, Wan B, Lei G, Tian X, Guo Y, Zhu J. Multiobjective and multiphysics design optimization of a switched reluctance motor for electric vehicle applications. IEEE Trans Energy Convers. 2021;36(4):3294–304. https://doi.org/10.1109/TEC.2021.3078547.
24. Ilhan E, Kremers MFJ, Motoasca ET, Paulides JJH, Lomonova EA. Spatial discretization methods for air gap permeance calculations in double salient traction motors. IEEE Trans Ind Appl. 2012;48(6):2165–72. https://doi.org/10.1109/TIA.2012.2226692.
25. Golzarzadeh M, Ganji B. Analytical modelling of the linear switched reluctance motor with segmental translator. IET Electr Power Appl. 2019;13(4):527–37. https://doi.org/10.1049/iet-epa.2018.5516.
26. Djelloul-Khedda Z, Boughrara K, Dubas F, Ibtiouen R. Nonlinear analytical prediction of magnetic field and electromagnetic performances in switched reluctance machines. IEEE Trans Magn. 2017;53(7):1–11. https://doi.org/10.1109/TMAG.2017.2679686.
27. Song S, Zhang M, Ge L. A new decoupled analytical modeling method for switched reluctance machine. IEEE Trans Magn. 2015;51(3):1–4. https://doi.org/10.1109/TMAG.2014.2363214.
28. Xia CL, Xue M, Shi TN. A new rapid nonlinear simulation method for switched reluctance motors. IEEE Trans Energy Convers. 2009;24(3):578–86. https://doi.org/10.1109/TEC.2009.2016131.
29. Cai Y, Wang Y, Xu H, Sun S, Wang C, Sun L. Research on rotor position model for switched reluctance motor using neural network. IEEE/ASME Trans Mechatron. 2018;23(6):2762–73. https://doi.org/10.1109/TMECH.2018.2870892.
30. Dong J, Howey B, Danen B, Lin J, Jiang JW, Bilgin B, et al. Advanced dynamic modeling of three-phase mutually coupled switched reluctance machine. IEEE Trans Energy Convers. 2018;33(1):146–54. https://doi.org/10.1109/TEC.2017.2724765.
31. Sun W, Li Q, Sun L, Zhu L, Li L. Electromagnetic analysis on novel rotor-segmented axial-field SRM based on dynamic magnetic equivalent circuit. IEEE Trans Magn. 2019;55(6):1–5. https://doi.org/10.1109/TMAG.2019.2901002.
32. Li S, Zhang S, Dang J, Habetler TG, Harley RG. Analytical calculation of the phase inductance profile of switched reluctance machines. IEEE Trans Energy Convers. 2019;34(3):1149–63. https://doi.org/10.1109/TEC.2019.2890821.

33. Yu Q, Wang X, Cheng Y. Magnetic modeling of saliency effect for saturated electrical machines with a new calculation method. IEEE Trans Magn. 2016;52(6):1–6. https://doi.org/10.1109/TMAG.2016.2521344.
34. Shang-Hsun M, Mi-Ching T. A novel switched reluctance motor with C-core stators. IEEE Trans Magn. 2005;41(12):4413–20. https://doi.org/10.1109/tmag.2005.858372.
35. Ding W, Hu Y, Wu L. Analysis and development of novel three-phase hybrid magnetic paths switched reluctance motors using modular and segmental structures for EV applications. IEEE/ASME Trans Mechatron. 2015;20(5):2437–51. https://doi.org/10.1109/tmech.2014.2383615.
36. Jalali Kondelaji MA, Farahani EF, Mirsalim M. Teethed-pole switched reluctance motors assisted with permanent magnets: analysis and evaluation. IEEE Trans Energy Convers. 2021;36(3):2131–40. https://doi.org/10.1109/tec.2021.3055739.
37. Cheewoo L, Krishnan R. New designs of a two-phase E-core switched reluctance machine by optimizing the magnetic structure for a specific application: concept, design, and analysis. IEEE Trans Ind Appl. 2009;45(5):1804–14. https://doi.org/10.1109/tia.2009.2027570.
38. Lee C, Krishnan R, Lobo NS. Novel two-phase switched reluctance machine using common-pole E-core structure: concept, analysis, and experimental verification. IEEE Trans Ind Appl. 2009;45(2):703–11. https://doi.org/10.1109/tia.2009.2013592.
39. Eskandari H, Mirsalim M. An improved 9/12 two-phase E-core switched reluctance machine. IEEE Trans Energy Convers. 2013;28(4):951–8. https://doi.org/10.1109/tec.2013.2279344.
40. Mousavi-Aghdam SR, Feyzi MR, Bianchi N, Morandin M. Design and analysis of a novel high-torque stator-segmented SRM. IEEE Trans Ind Electron. 2016;63(3):1458–66. https://doi.org/10.1109/tie.2015.2494531.
41. Mecrow BC, Finch JW, El-Kharashi EA, Jack AG. Switched reluctance motors with segmental rotors. IEE Proc Electr Power Appl. 2002;149(4):245–54. https://doi.org/10.1049/ip-epa:20020345.
42. Mecrow BC, El-Kharashi EA, Finch JW, Jack AG. Segmental rotor switched reluctance motors with single-tooth windings. IEE Proc Electr Power Appl. 2003;150(5):591–9. https://doi.org/10.1049/ip-epa:20030366.
43. Mecrow BC, El-Kharashi EA, Finch JW, Jack AG. Preliminary performance evaluation of switched reluctance motors with segmental rotors. IEEE Trans Energy Convers. 2004;19(4):679–86. https://doi.org/10.1109/TEC.2004.837290.
44. Sun X, Diao K, Lei G, Guo Y, Zhu J. Study on segmented-rotor switched reluctance motors with different rotor pole numbers for BSG system of hybrid electric vehicles. IEEE Trans Veh Technol. 2019;68(6):5537–47. https://doi.org/10.1109/TVT.2019.2913279.
45. Ding W, Yang S, Hu Y, Li S, Wang T, Yin Z. Design consideration and evaluation of a 12/8 high-torque modular-stator hybrid excitation switched reluctance machine for EV applications. IEEE Trans Ind Electron. 2017;64(12):9221–32. https://doi.org/10.1109/tie.2017.2711574.
46. Ding W, Yang S, Hu Y. Development and investigation on segmented-stator hybrid-excitation switched reluctance machines with different rotor pole numbers. IEEE Trans Ind Electron. 2018;65(5):3784–94. https://doi.org/10.1109/tie.2017.2760846.
47. Masoumi M, Mirsalim M. E-core hybrid reluctance motor with permanent magnets inside stator common poles. IEEE Trans Energy Convers. 2018;33(2):826–33. https://doi.org/10.1109/tec.2017.2769621.
48. Farahani EF, Kondelaji MAJ, Mirsalim M. An innovative hybrid-excited multi-tooth switched reluctance motor for torque enhancement. IEEE Trans Ind Electron. 2021;68(2):982–92. https://doi.org/10.1109/tie.2020.2969073.
49. Abbasian M, Moallem M, Fahimi B. Double-stator switched reluctance machines (DSSRM): fundamentals and magnetic force analysis. IEEE Trans Energy Convers. 2010;25(3):589–97. https://doi.org/10.1109/tec.2010.2051547.
50. Isfahani AH, Fahimi B. Comparison of mechanical vibration between a double-stator switched reluctance machine and a conventional switched reluctance machine. IEEE Trans Magn. 2014;50(2):293–6. https://doi.org/10.1109/tmag.2013.2286569.
51. Arbab N, Wang W, Lin C, Hearron J, Fahimi B. Thermal modeling and analysis of a double-stator switched reluctance motor. IEEE Trans Energy Convers. 2015;30(3):1209–17. https://doi.org/10.1109/tec.2015.2424400.

52. Asgar M, Afjei E, Torkaman H. A new strategy for design and analysis of a double-stator switched reluctance motor: electromagnetics, fem, and experiment. IEEE Trans Magn. 2015;51(12):1–8. https://doi.org/10.1109/tmag.2015.2465307.
53. Lee CHT, Chau KT, Liu C, Ching TW, Li F. Mechanical offset for torque ripple reduction for magnetless double-stator doubly salient machine. IEEE Trans Magn. 2014;50(11):1–4. https://doi.org/10.1109/tmag.2014.2320964.
54. Asgar M, Afjei E. Radial force reduction in a new flat-type double-stator switched reluctance motor. IEEE Trans Energy Convers. 2016;31(1):141–9. https://doi.org/10.1109/tec.2015.2465833.
55. Gupta TD, Chaudhary K, Elavarasan RM, Saket RK, Khan I, Hossain E. Design modification in single-tooth winding double-stator switched reluctance motor for torque ripple mitigation. IEEE Access. 2021;9:19078–96. https://doi.org/10.1109/access.2021.3052828.
56. Shumei C, Yongjie Y, Tiecheng W. Research on switched reluctance double-rotor motor used for hybrid electric vehicle. In: Proceedings international conference on electrical machines and systems. 2008. p. 3393–6.
57. Yoshimaru Y, Higuchi T, Yokoi Y, Abe T. On the improvement performance of a dual rotor segment type SRM. In: International conference on power electronics drive systems. 2013. p. 1045–8.
58. Guo T, Schofield N, Emadi A. Double segmented rotor switched reluctance machine with shared stator back-iron for magnetic flux passage. IEEE Trans Energy Convers. 2016;31(4):1278–86. https://doi.org/10.1109/tec.2016.2600178.
59. Yang Y, Schofield N, Emadi A. Double-rotor switched reluctance machine (DRSRM). IEEE Trans Energy Convers. 2015;30(2):671–80. https://doi.org/10.1109/tec.2014.2378211.
60. Yang H-Y, Lim Y-C, Kim H-C. Acoustic noise/vibration reduction of a single-phase SRM using skewed stator and rotor. IEEE Trans Ind Electron. 2013;60(10):4292–300. https://doi.org/10.1109/tie.2012.2217715.
61. Gan C, Wu J, Shen M, Yang S, Hu Y, Cao W. Investigation of skewing effects on the vibration reduction of three-phase switched reluctance motors. IEEE Trans Magn. 2015;51(9):1–9. https://doi.org/10.1109/tmag.2015.2441035.
62. Yasa Y, Elamin M, Sozer Y, Kutz J, Tylenda JS, Wright RL. Acoustic noise mitigation for high pole count switched reluctance machines through skewing method with multiphysics FEA simulations. In: Proceedings IEEE energy conversion congress and exposition. 2017. p. 738–44
63. Torkaman H, Ghaheri A, Keyhani A. Axial flux switched reluctance machines: a comprehensive review of design and topologies. IET Electr Power Appl. 2019;13(3):310–21.
64. Arihara H, Akatsu K. Basic properties of an axial-type switched reluctance motor. IEEE Trans Ind Appl. 2013;49(1):59–65. https://doi.org/10.1109/tia.2012.2229683.
65. Labak A, Kar NC. Designing and prototyping a novel five-phase pancake-shaped axial-flux SRM for electric vehicle application through dynamic FEA incorporating flux-tube modeling. IEEE Trans Ind Appl. 2013;49(3):1276–88. https://doi.org/10.1109/tia.2013.2252871.
66. Wang D, Du X, Zhang D, Wang X. Design, optimization, and prototyping of segmental-type linear switched-reluctance motor with a toroidally wound mover for vertical propulsion application. IEEE Trans Ind Electron. 2018;65(2):1865–74. https://doi.org/10.1109/TIE.2017.2740824.
67. Byeong-Seok L, Han-Kyung B, Vijayraghavan P, Krishnan R. Design of a linear switched reluctance machine. IEEE Trans Ind Appl. 2000;36(6):1571–80. https://doi.org/10.1109/28.887208.
68. Lobo NS, Lim HS, Krishnan R. Comparison of linear switched reluctance machines for vertical propulsion application: analysis, design, and experimental correlation. IEEE Trans Ind Appl. 2008;44(4):1134–42. https://doi.org/10.1109/TIA.2008.926294.
69. Cheng-Tsung L, Kun-Shian S, Jyh-Wei C. Operational stability enhancement analysis of a transverse flux linear switched-reluctance motor. IEEE Trans Magn. 2000;36(5):3699–702. https://doi.org/10.1109/20.908945.

70. Baoming G, de Almeida AT, Ferreira FJTE. Design of transverse flux linear switched reluctance motor. IEEE Trans Magn. 2009;45(1):113–9. https://doi.org/10.1109/TMAG.2008.2006193
71. Wang D, Zhang D, Du X, Wang X. Unitized design methodology of linear switched reluctance motor with segmental secondary for long rail propulsion application. IEEE Trans Ind Electron. 2018;65(12):9884–94. https://doi.org/10.1109/TIE.2018.2829690.
72. Ganji B, Askari MH. Different topologies for linear switched reluctance motor with segmental translator. In: Proceedings 24th Iranian conference on electrical engineering. 2016. p. 874–8.
73. Szabo L, Ruba M. Segmental stator switched reluctance machine for safety-critical applications. IEEE Trans Ind Appl. 2012;48(6):2223–9. https://doi.org/10.1109/tia.2012.2226857.
74. Widmer JD, Mecrow BC. Optimized segmental rotor switched reluctance machines with a greater number of rotor segments than stator slots. IEEE Trans Ind Appl. 2013;49(4):1491–8. https://doi.org/10.1109/tia.2013.2255574.
75. Widmer JD, Martin R, Mecrow BC. Optimization of an 80-kW segmental rotor switched reluctance machine for automotive traction. IEEE Trans Ind Appl. 2015;51(4):2990–9. https://doi.org/10.1109/TIA.2015.2405051.
76. Sun X, Feng L, Diao K, Yang Z. An improved direct instantaneous torque control based on adaptive terminal sliding mode for a segmented-rotor SRM. IEEE Trans Ind Electron. 2021;68(11):10569–79. https://doi.org/10.1109/TIE.2020.3029463.
77. Zhu J, Cheng KWE, Xue X. Design and analysis of a new enhanced torque hybrid switched reluctance motor. IEEE Trans Energy Convers. 2018;33(4):1965–77. https://doi.org/10.1109/tec.2018.2876306.
78. Lu K, Jakobsen U, Rasmussen PO. Single-phase hybrid switched reluctance motor for low-power low-cost applications. IEEE Trans Magn. 2011;47(10):3288–91. https://doi.org/10.1109/tmag.2011.2157466.
79. Zou Y, Cheng KWE, Cheung NC, Pan J. Deformation and noise mitigation for the linear switched reluctance motor with skewed teeth structure. IEEE Trans Magn. 2014;50(11):1–4. https://doi.org/10.1109/tmag.2014.2323420
80. Labak A, Kar NC. Novel approaches towards leakage flux reduction in axial flux switched reluctance machines. IEEE Trans Magn. 2013;49(8):4738–41. https://doi.org/10.1109/TMAG.2013.2261287.
81. Ebrahimi Y, Feyzi MR. Introductory assessment of a novel high-torque density axial flux switched reluctance machine. IET Electr Power Appl. 2017;11(7):1315–23. https://doi.org/10.1049/iet-epa.2017.0081.
82. Belhadi MH, Krebs G, Marchand C, Hannoun H, Mininger X. Evaluation of axial SRM for electric vehicle application. Electr Power Syst Res. 2017;148:155–61. https://doi.org/10.1016/j.epsr.2017.03.034.
83. Wang D, Wang X, Du X. Design and comparison of a high force density dual-side linear switched reluctance motor for long rail propulsion application with low cost. IEEE Trans Magn. 2017;53(6):1–4. https://doi.org/10.1109/TMAG.2017.2659804.
84. Pan JF, Zou Y, Cheung N, Cao G. On the voltage ripple reduction control of the linear switched reluctance generator for wave energy utilization. IEEE Trans Power Electron. 2014;29(10):5298–307. https://doi.org/10.1109/TPEL.2013.2292069.
85. Lim HS, Krishnan R, Lobo NS. Design and control of a linear propulsion system for an elevator using linear switched reluctance motor drives. IEEE Trans Ind Electron. 2008;55(2):534–42. https://doi.org/10.1109/TIE.2007.911942.
86. Han-Kyung B, Byeong-Seok L, Vijayraghavan P, Krishnan R. A linear switched reluctance motor: converter and control. IEEE Trans Ind Appl. 2000;36(5):1351–9. https://doi.org/10.1109/28.871284.

Chapter 3
Design Methods for Switched Reluctance Motors

Abstract The flux linkage characteristics of switched reluctance motors (SRMs) demonstrate a highly nonlinear correlation with phase current and rotor position. This inherent attribute introduces numerous uncertainties in the design process of SRMs, with many dimensions primarily determined through the utilization of empirical formulas during the initial stages. These formulas incorporate certain assumptions regarding the characteristics of SRMs. In this chapter, the conventional design method is introduced first. This approach relies on empirical rules based on the designer's experience. Then, a novel magnetic parameter design method is proposed to overcome the conventional approach's limitations. Each symbol in this method has a comprehensible physical meaning. The proposed method can effectively provide ideal design results for SRMs and eliminate the meaningless empirical coefficients of the conventional method. Finally, design examples are investigated.

Keywords Design method · Nonlinear characteristic · Switched reluctance motors

3.1 Introduction

Switched reluctance motors (SRMs) have gained significant attention in recent years due to their numerous advantages over conventional motors. The unique construction and operational principles of SRMs make them suitable for various applications, including electric vehicles, industrial automation, and renewable energy systems [1, 2]. To harness the full potential of SRMs, it is essential to employ effective design methods that optimize their performance, efficiency, and reliability.

The design of SRMs poses unique challenges compared to conventional motor types. The magnetic characteristics and performance of SRMs are difficult to predict since the flux linkage exhibits nonlinear relations concerning the phase current and rotor position [3–6]. There are two main reasons for this nonlinearity. First, for the high requirement of torque density, SRMs commonly operate in the magnetic saturation region, which results in a nonlinear relationship between flux linkage and current. The other is that the reluctance between the stator and rotor varies significantly due

X. Sun et al., *Multi-objective Design Optimization of Switched Reluctance Motor Drive Systems*, https://doi.org/10.1007/978-981-96-0672-6_3

to the double salient structure of SRMs [7–10]. This inherent characteristic yields many uncertainties in the design of SRMs.

In general, an SRM is initially designed using the analytical method. The foundational mathematical design theory of SRMs was presented in [11], and more papers investigated the effect of the number of stator/rotor poles on the performance of SRMs [12–15]. In [16], a design flowchart based on the analytical method was presented for the SRMs. The lamination geometry was determined by the flux splitting ratios, and the number of turns was calculated by the simple peak flux equation.

This chapter will focus on the design methods for SRMs, including the principles and derivations. The remainder of this chapter is organized as follows. Section 3.2 provides the conventional design method for SRMs. Section 3.3 introduces a novel magnetic parameter design method that can greatly reduce the overdependence of the designer's experience. Section 3.4 investigates design examples to verify the proposed novel design method, followed by the summary in Sect. 3.5.

3.2 The Conventional Design Method

The electromagnetic load of a motor refers to its electrical and magnetic loads. They are directly related to the determination of the main dimensions of the motor and affect the operating characteristics, efficiency, temperature rise, and other important parameters in motor design.

The electric load A refers to the total current in the conductor per unit length on the armature's peripheral surface, with the SRM armature's diameter equal to the inner diameter of the stator. Due to the small air gap, the rotor diameter can be used instead. Thus,

$$A = \frac{mN_{\text{ph}}I}{\pi D_{\text{si}}} \approx \frac{mN_{\text{ph}}I}{\pi D} \tag{3.1}$$

where m is the phase number, N_{ph} is the number of turns per phase, I is the root mean square (RMS) value of the winding current, and D_{si} and D are the stator inner diameter and rotor outer diameter, respectively.

The RMS value of the winding current can be expressed as:

$$I = \sqrt{\frac{1}{T}\int_{0}^{T} i^2 dt} = k_{\text{i}} i_{\text{m}} \tag{3.2}$$

where T is the time of one current cycle, i_{m} is the actual peak current value, and k_{i} is the peak current coefficient related to the current waveform.

Most SRMs use the angular position control (APC) method to operate at their rated operating point. To achieve a high operating efficiency, the current waveform given by the APC method should be close to a flat-top wave. Therefore, the current at the rated operating point of the SRM can be assumed as a square wave with an amplitude I_m. The equivalent amplitude of the square wave current I_m differs from the actual current amplitude i_m. The equivalent condition is that the effective values of the two currents are the same.

Assume that $I_m = k_m i_m$, thus

$$I_m = \frac{k_m}{k_i} I. \tag{3.3}$$

Since each phase winding of the SRM has two states, conduction and non-conduction, the load factor K_d is defined as

$$K_d = \frac{\theta_{off} - \theta_{on}}{2\pi / N_r} = \frac{N_r}{2\pi}\theta_c \tag{3.4}$$

where θ_{on}, θ_{off}, and θ_c are the turn-on, turn-off, and conduction angles, respectively, and N_r is the number of rotor poles.

Under the assumption of square wave current and neglecting the stator winding resistance, the electromagnetic power can be expressed as

$$P_{em} = mUI_m K_d \tag{3.5}$$

where U represents the rated voltage of the armature coil.

According to the voltage balance equation, we have

$$\Psi_m = \frac{U}{\Omega}(\theta_{off} - \theta_{on}) = \frac{U}{\Omega}\theta_c \tag{3.6}$$

and the rated rotor angular speed in rad/s,

$$\Omega = \frac{2\pi n}{60} \tag{3.7}$$

where n represents the motor rotational speed in revolutions per minute or rev/min.

The maximum flux Ψ_m occurs at the turn-off angular position. The magnetic load of the motor refers to the average magnetic flux density in the air gap. Due to the non-uniform magnetic circuit, it is difficult to describe the magnetic load of SRM motors. When a single-phase winding of the SRM is energized, it can be seen from the magnetic field distribution that the main flux in each pole is within a range of the rotor pole pitch. Thus, the magnetic load can be defined as

$$B_\delta = \frac{\Phi}{\tau_r l} \tag{3.8}$$

where Φ is the main flux of each pole at position θ_{off}, l is the axial length, and τ_r is the rotor pole pitch, which is expressed as

$$\tau_r = \frac{D\pi}{N_r}. \tag{3.9}$$

After determining the magnetic load, the maximum flux linkage can be expressed as a function of B_δ, as follows

$$\Psi_m = N_{ph}\Phi = N_{ph}B_\delta Dl\frac{\pi}{N_r}. \tag{3.10}$$

According to the above equations, we have

$$D^2 l = \frac{6.1}{B_\delta A}\cdot\frac{k_i}{k_m}\frac{P_{em}}{n}. \tag{3.11}$$

The following conclusions can be drawn from (3.11):

(1) When the electromagnetic load A and B_δ are constant, the volume of the SRM is proportional to the electromagnetic power and inversely proportional to the rated speed. Since P_{em}/n is proportional to the rated torque, and $D^2 l$ depends on the rated torque value. On the other hand, when the electromagnetic power is fixed, the lower the speed, the smaller the volume.
(2) For an SRM with a given power and speed, both the size and material will decrease with the increase of the electromagnetic load A and B_δ. However, the selection of high values of A and B_δ should consider the temperature rise of the motor.
(3) Considering the value of k_i/k_m, the current waveform has an impact on the volume of SRM. If the SRM is dimensioned with an ideal square wave current, $k_i/k_m = 1/\sqrt{2}$.

Define the ratio of axial length to rotor outer diameter and the ratio of stator outer diameter to rotor outer diameter as

$$\lambda = l/D, \tag{3.12}$$

$$\varepsilon = D/D_{so}. \tag{3.13}$$

Commonly, the reasonable value ranges of λ and ε are 0.5–3.0 and 0.4–0.7, respectively. Based on the Eqs. (3.11)–(3.13), the main dimensions of the motor can be achieved.

Combining (3.6), (3.7), and (3.10), the number of turns per phase N_{ph} can be expressed as

$$N_{ph} = \frac{N_r U\theta_c}{\pi\Omega B_\delta Dl} = \frac{30}{\pi^2}\frac{N_r U\theta_c}{nB_\delta Dl} \approx \frac{3.04N_r U\theta_c}{nB_\delta Dl}. \tag{3.14}$$

3.3 Novel Magnetic Parameter Design Method

Figure 3.1 shows an example of flux curves in a single stator tooth of SRMs. The degrees 0° and 180° in the figure are the electric degrees, which represent the unaligned and aligned positions, respectively. The inflexion point at the 180° line is denoted as A. The flux linkage and current are regarded as Ψ_s and I_s. The point at 0° line with the same current of A is denoted as B, and its corresponding flux linkage is Ψ_b. Since the motor should exhibit a certain overload capacity, the SRM operates in startup mode during design. The chopping current is set as I_H, and corresponding points at 180° and 0° lines are denoted as C and D, respectively. Ψ_H and Ψ_d are the flux linkage values of C and D, respectively [17].

The whole area of SRM under current chopping control (CCC) can be divided into the unsaturated area ($I < I_s$) and saturated area ($I_s < I < I_H$), as presented in Fig. 3.1. To meet certain overload capacity, point A is selected as the rated operating point at the startup mode for the design of the SRM. All the following design rules are based on this assumed condition.

3.3.1 *Design of Structural Parameters*

The magnetic energy ΔW can be calculated by the area of AODCA shown in Fig. 3.1. Assume the flux slope at 0° zero and ignore the flux slope of line AC. The area AODCA can be estimated as a trapezoid with top base $I_H - I_s$, bottom base I_H, and height Ψ_H. Thus, ΔW is estimated by

$$\Delta W \approx \frac{1}{2}\psi_s[(I_H - I_s) + I_H] = \psi_s\left(I_H - \frac{1}{2}I_s\right). \tag{3.15}$$

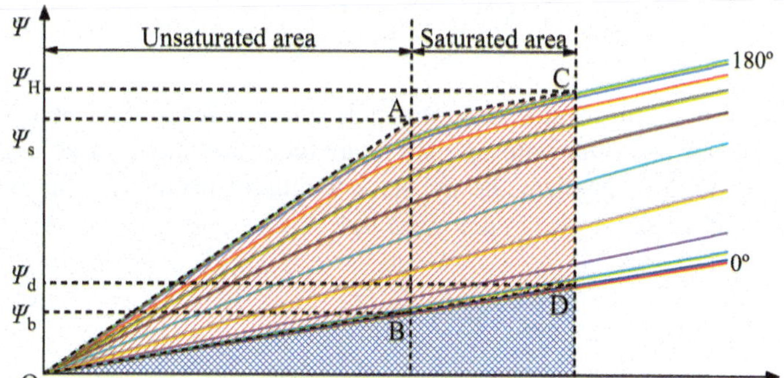

Fig. 3.1 Flux curves in a single stator tooth of the SRM in terms of electric degrees

It should be noted that there exists an unavoidable error in (3.15) when the flux slope at 0° is not close to 0. Thus, it should be further modified. Define $K_L = L_{max}/L_{min} = \Psi_s/\Psi_d$, in which L_{max} and L_{min} mean the maximum and minimum inductances, respectively. It can be found that K_L is a constant when the SRM operates at point A or within the unsaturated area. The flux linkage at point D can be represented by

$$\psi_d = \frac{I_H}{I_s}\psi_b = \frac{I_H}{I_s}\frac{\psi_s}{K_L}. \tag{3.16}$$

Therefore, by subtracting the area below the flux line of 0°, (3.15) can be modified as

$$\Delta W \approx \psi_s\left(I_H - \frac{1}{2}I_s\right) - \frac{1}{2}I_H\psi_d = \psi_s\left(I_H - \frac{I_s}{2} - \frac{I_H^2}{2I_sK_L}\right). \tag{3.17}$$

Assume that the conduction angle θ_m is half of the period for the SRM, we have

$$\theta_m = \frac{\pi}{N_r}. \tag{3.18}$$

Then, the torque generated by the single stator tooth can be expressed as

$$T_{pole} = \frac{\Delta W}{\Delta\theta} = \frac{\Delta W}{\theta_m} = \frac{\psi_s\left(I_H - \frac{I_s}{2} - \frac{I_H^2}{2K_LI_s}\right)}{\pi/N_r}. \tag{3.19}$$

The air gap surface area corresponding to each stator pole can be calculated by

$$S = \begin{cases} \frac{\alpha_s\pi Dl}{N_s} & \beta_s < \beta_r \\ \frac{\alpha_r\pi Dl}{N_r} & \beta_s > \beta_r \end{cases} \tag{3.20}$$

where α_s and α_r are the stator and rotor pole arc coefficients, N_s and N_r are the numbers of stator and rotor poles, β_s and β_r are the stator and rotor pole arc, and D and l are the stator inner diameter and stack length, respectively. The relationship between α_s and β_s, and α_r and β_r are given by

$$\alpha_s = \frac{\beta_s}{360°/N_s}, \tag{3.21}$$

$$\alpha_r = \frac{\beta_r}{360°/N_r}. \tag{3.22}$$

Therefore, the magnetic flux can be expressed as

$$\phi_s = B_s S = \begin{cases} B_s \frac{\alpha_s \pi Dl}{N_s} & \beta_s < \beta_r \\ B_s \frac{\alpha_r \pi Dl}{N_r} & \beta_s > \beta_r \end{cases} \tag{3.23}$$

where B_s is the flux density at the rated operation point.

Ignoring the flux leakage, the flux linkage of a single tooth is

$$\psi_s = N_{\text{pole}} \phi_s = \begin{cases} \frac{N_{\text{pole}} B_s \alpha_s \pi Dl}{N_s} & \beta_s < \beta_r \\ \frac{N_{\text{pole}} B_s \alpha_r \pi Dl}{N_r} & \beta_s > \beta_r \end{cases} \tag{3.24}$$

where N_{pole} is the number of turns per stator pole.

Since the total number of stator poles is N_s, and each pole is excited under half of the period, the total torque generated by all the stator poles is

$$T_L = N_s T_{\text{pole}}/2. \tag{3.25}$$

Substituting (3.19) and (3.24) into (3.25), we have

$$T_L = \begin{cases} \frac{N_r N_{\text{pole}} B_s \alpha_s Dl}{2} \left(I_H - \frac{I_s}{2} - \frac{I_H^2}{2K_L I_s} \right) & \beta_s < \beta_r \\ \frac{N_s N_{\text{pole}} B_s \alpha_r Dl}{2} \left(I_H - \frac{I_s}{2} - \frac{I_H^2}{2K_L I_s} \right) & \beta_s > \beta_r \end{cases}. \tag{3.26}$$

The unknown parameters in the Eq. (3.26) are I_s and I_H. Thus, the next step is to replace the current parameters with the specific electric load. According to (3.1), the specific electric load can be determined by

$$A = \frac{mN_{\text{ph}} I}{\pi D_{\text{si}}} \approx \frac{mN_{\text{ph}} I}{\pi D} = \frac{2N_s N_{\text{pole}} I}{\pi D} \tag{3.27}$$

where I is the RMS value of the current.

Define the rated specific electric load at point A as A_s and the maximum specific electric load at point C as A_H, and they can be calculated by

$$A_s = \frac{2N_s N_{\text{pole}} I_s}{\sqrt{2} \pi D}, \tag{3.28}$$

$$A_H = \frac{2N_s N_{\text{pole}} I_H}{\sqrt{2} \pi D}. \tag{3.29}$$

Further, I_s and I_H can be expressed by

$$I_s = \frac{\pi D A_s}{\sqrt{2} N_s N_{\text{pole}}}, \tag{3.30}$$

$$I_{\mathrm{H}} = \frac{\pi D A_{\mathrm{H}}}{\sqrt{2} N_{\mathrm{s}} N_{\mathrm{pole}}}. \tag{3.31}$$

Substituting (3.30) and (3.31) into (3.26), we have

$$T_{\mathrm{L}} = \begin{cases} \frac{\pi N_{\mathrm{r}} B_{\mathrm{s}} \alpha_{\mathrm{s}} D^2 l}{2\sqrt{2} N_{\mathrm{s}}} \left(A_{\mathrm{H}} - \frac{A_{\mathrm{s}}}{2} - \frac{A_{\mathrm{H}}^2}{2 K_{\mathrm{L}} A_{\mathrm{s}}} \right) & \beta_{\mathrm{s}} < \beta_{\mathrm{r}} \\ \frac{\pi B_{\mathrm{s}} \alpha_{\mathrm{r}} D^2 l}{2\sqrt{2}} \left(A_{\mathrm{H}} - \frac{A_{\mathrm{s}}}{2} - \frac{A_{\mathrm{H}}^2}{2 K_{\mathrm{L}} A_{\mathrm{s}}} \right) & \beta_{\mathrm{s}} > \beta_{\mathrm{r}} \end{cases}. \tag{3.32}$$

The relationship between the rated torque T and the starting torque T_{L} is

$$T = T_{\mathrm{L}} / k_{\mathrm{n}} \tag{3.33}$$

where k_{n} is the overload factor.

Therefore, the overall dimensions of the SRM can be determined by

$$D^2 l = \begin{cases} \frac{2\sqrt{2} k_{\mathrm{n}} N_{\mathrm{s}} T}{\pi N_{\mathrm{r}} B_{\mathrm{s}} \alpha_{\mathrm{s}} \left(A_{\mathrm{H}} - \frac{A_{\mathrm{s}}}{2} - \frac{A_{\mathrm{H}}^2}{2 K_{\mathrm{L}} A_{\mathrm{s}}} \right)} & \beta_{\mathrm{s}} < \beta_{\mathrm{r}} \\ \frac{2\sqrt{2} k_{\mathrm{n}} T}{\pi B_{\mathrm{s}} \alpha_{\mathrm{r}} \left(A_{\mathrm{H}} - \frac{A_{\mathrm{s}}}{2} - \frac{A_{\mathrm{H}}^2}{2 K_{\mathrm{L}} A_{\mathrm{s}}} \right)} & \beta_{\mathrm{s}} > \beta_{\mathrm{r}} \end{cases}. \tag{3.34}$$

Then, the relationship between D and l, and the relationship between D and D_{so} can be found from (3.12) and (3.13), respectively.

3.3.2 *Design of Winding Parameters*

Based on (3.24) and (3.18), the electromotive force of each pole can be expressed as

$$E_{\mathrm{pole}} = \frac{\Delta \psi}{\Delta t} = \omega \frac{\psi_{\mathrm{s}}}{\theta_{\mathrm{m}}} = \begin{cases} \frac{\pi}{30} n B_{\mathrm{s}} \frac{\alpha_{\mathrm{s}} N_{\mathrm{r}} D l}{N_{\mathrm{s}}} N_{\mathrm{pole}} & \beta_{\mathrm{s}} < \beta_{\mathrm{r}} \\ \frac{\pi}{30} n B_{\mathrm{s}} \alpha_{\mathrm{r}} D l N_{\mathrm{pole}} & \beta_{\mathrm{s}} > \beta_{\mathrm{r}} \end{cases} \tag{3.35}$$

where ω and n are the rated angular speed in rad/s and the rotating speed in rev/min of the SRM, respectively. The relationship between the electromotive force of each phase and each pole can be represented by

$$E_{\mathrm{ph}} = N_{\mathrm{s}} E_{\mathrm{pole}} / n_{\mathrm{p}} m \tag{3.36}$$

where n_{p} is the number of parallel branches.

Besides, the electromotive force of each phase is given by

$$E_{ph} = k_e U \tag{3.37}$$

where k_e is the potential coefficient in the range of 0.8–0.9 and U is the rated voltage.

According to (3.35), (3.36) and (3.37), the number of turns each pole can be determined by

$$N_{pole} = \begin{cases} \frac{30n_p m k_e U}{\pi n B_s \alpha_s N_r D l} & \beta_s < \beta_r \\ \frac{30n_p m k_e U}{\pi n B_s \alpha_r N_s D l} & \beta_s > \beta_r \end{cases} \tag{3.38}$$

3.4 Design Procedure and Case Study

This section presents a generic design procedure for SRMs by using the discussed design methods. A case study with some experimental results is also investigated to show the effectiveness of the proposed procedure. As the prototype is obtained from optimization, a multi-objective optimization method is also included in the procedure. The optimization is necessary for the procedure as it will ensure a fair comparison for different motor topologies, i.e., the comparison is carried out based on the optimal design for each motor option. Therefore, three SRMs with multiple phases are initially designed and further optimized for performance improvement in this section. Besides, the proposed magnetic parameter design methodology will be verified by the optimized SRMs.

Figure 3.2 shows the flowchart of the design optimization process.

3.4.1 *Initial Design*

In this part, three SRMs, i.e., a three-phase 12/8 SRM, a five-phase 10/8 SRM, and a nine-phase 18–20 SRM, are initially designed based on the proposed magnetic parameter design methodology. The relationship between m, N_s, and N_r should meet [14]

$$\mathrm{LCM}(N_s, N_r) = mN_r \tag{3.39}$$

where LCM means the least common multiple.

The topologies of these SRMs are presented in Fig. 3.3.

The specifications for the SRMs are tabulated in Table 3.1. Take the three-phase SRM as an example and set $B_s = 1.3$ T, $\alpha_s = 0.45$, $\alpha_r = 0.35$, $K_L = 5$, $k_n = 02$, and $k_e = 0.8$. It can be calculated that $D^2 l = 1.4 \times 10^{-4}$ m^3 according to (3.34).

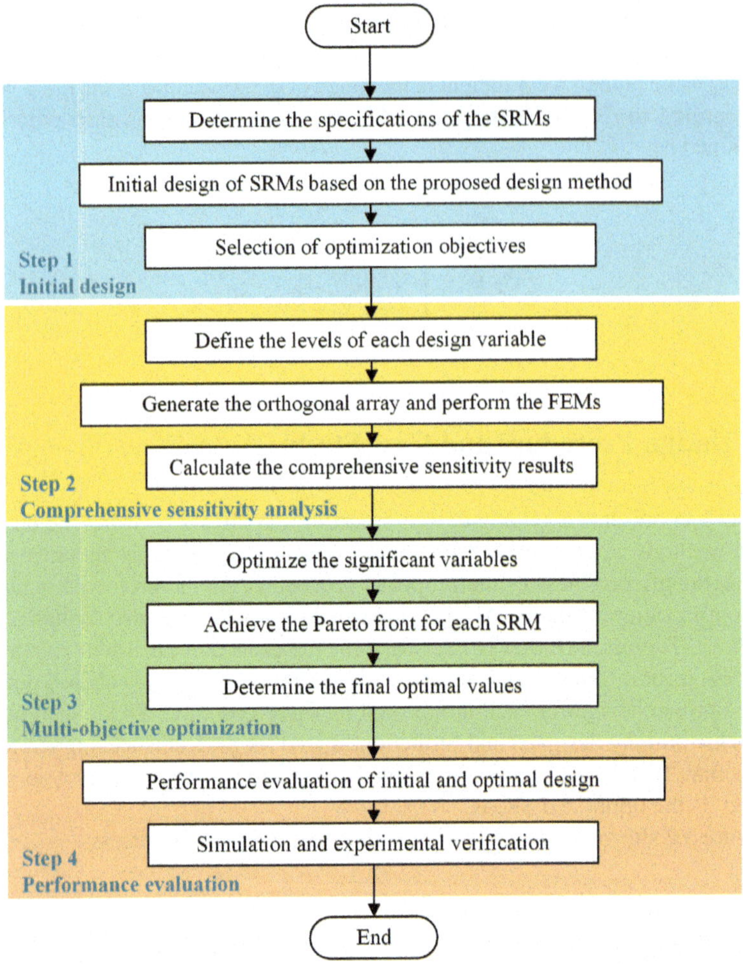

Fig. 3.2 A flowchart for the design optimization of SRMs

Set $\lambda = 1.4$, thus, $D = 46$ mm and $l = 65$ mm after calculation. Considering the stacking factor (set as 0.95 in this example) between adjacent silicon steel sheets, the actual stack length should be about 68 mm. The number of turns is calculated as 25 according to (3.35). For further analysis and comparison, the stack length of each motor is uniformly set as 70 mm. The initial setting and the calculated values of the main design variables are tabulated in Table 3.2.

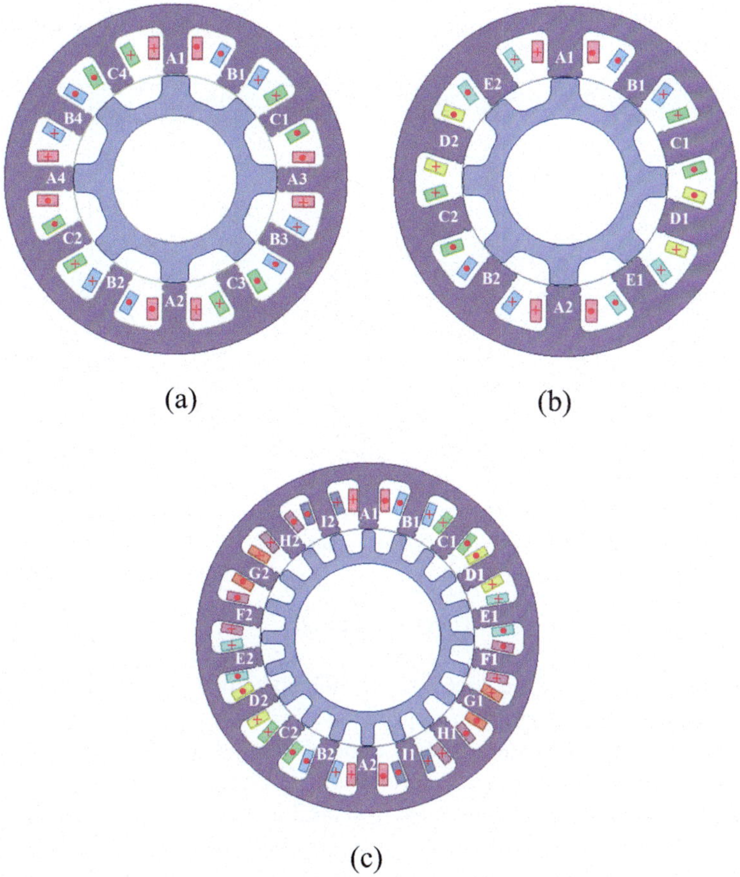

Fig. 3.3 Machine topologies for three SRMs. **a** Three-phase 12/8 SRM, **b** five-phase 10/8 SRM, and **c** nine-phase 18/20 SRM

Table 3.1 Specifications for the SRMs

Parameter	Unit	Value
Rated speed n	r/min	3000
Rated voltage U	V	24
Rated specific electric load A_s	A/mm	30,000
Maximum specific electric load A_H	A/mm	60,000

3.4.2 Multi-objective Optimization

The optimization problem can be defined as

Table 3.2 Initial setting and calculated design variable

Variable	12/8 SRM	10/8 SRM	18/20 SRM
m	3	5	9
N_s	12	10	18
N_r	8	8	20
K_L	5	4	3
n_p	2	1	1
λ	1.4	1.4	1.2
α_s	0.45	0.45	0.4
α_r	0.35	0.35	0.32
D (mm)	46	46	52
l (mm)	70	70	70
D_s (mm)	80	80	80
N_{pole}	25	21	21

$$\begin{cases} f_1(x_i) = \max\{T_{avg}\} \\ f_2(x_i) = \min\{J_c\} \\ f_3(x_i) = \min\{T_r\} \end{cases} \tag{3.40}$$

where T_{avg}, J_c and T_r are the three optimization objectives of average output torque, current density and torque ripple, respectively.

To simplify the optimization process, a comprehensive sensitivity analysis method is adopted to distinguish the significant and non-significant variables [18]. The design variables and their ranges are presented in Table 3.3, including four geometry parameters, i.e., D, α_s, α_r, and N_{pole}, and two control parameters, i.e., θ_{on} and θ_{off}. The sensitivity index of parameter x_i is calculated using the global sensitivity analysis method [2, 19].

Figure 3.4 shows the sensitivity analysis results for the three optimization objectives of the three SRMs. It can be noted that each variable presents different sensitivities to the optimization objectives. For example, in Fig. 3.4a, the number of turns per pole N_{pole} has a strong sensitivity to the output torque and current density but has little influence on torque ripple. The results of the turn-off angel θ_{off} are the opposite

Table 3.3 Design variables and their ranges

Variable	12/8 SRM	10/8 SRM	18/20 SRM
D (mm)	[45, 50]	[45, 50]	[50, 55]
α_s	[0.4, 0.5]	[0.4, 0.5]	[0.36, 0.46]
α_r	[0.3, 0.4]	[0.3, 0.4]	[0.26, 0.36]
N_{pole}	[21, 25]	[21, 25]	[21, 23]
θ_{on} (°)	[−2, 0]	[−2.4, 0]	[−1.4, 0]
θ_{off} (°)	[18, 20]	[17.6, 20]	[6.6, 8]

of N_{pole}. Therefore, selecting the key design variables based on sensitivity results is challenging. To solve this problem effectively, a comprehensive sensitivity function is adopted to quantitatively reflect the influence of each design variable, which is expressed as

$$S_{\text{total}}(x_i) = \omega_t S_t(x_i) + \omega_c S_c(x_i) + \omega_r S_r(x_i) \tag{3.41}$$

where $S_{\text{total}}(x_i)$ is the comprehensive sensitivity function, $S_t(x_i)$, $S_c(x_i)$, and $S_r(x_i)$ are the sensitivity values of output torque, current density, and torque ripple, respectively, and ω_t, ω_c, and ω_r are the weights of the three objectives, which can be determined by the design specifications. In this case study, the values of the three factors are assigned as 0.4, 0.3, and 0.3. The comprehensive sensitivity of each design variable is presented in Table 3.4. According to the calculated results, four variables, i.e., α_s, α_r, N, and θ_{on} are selected as the optimization variables.

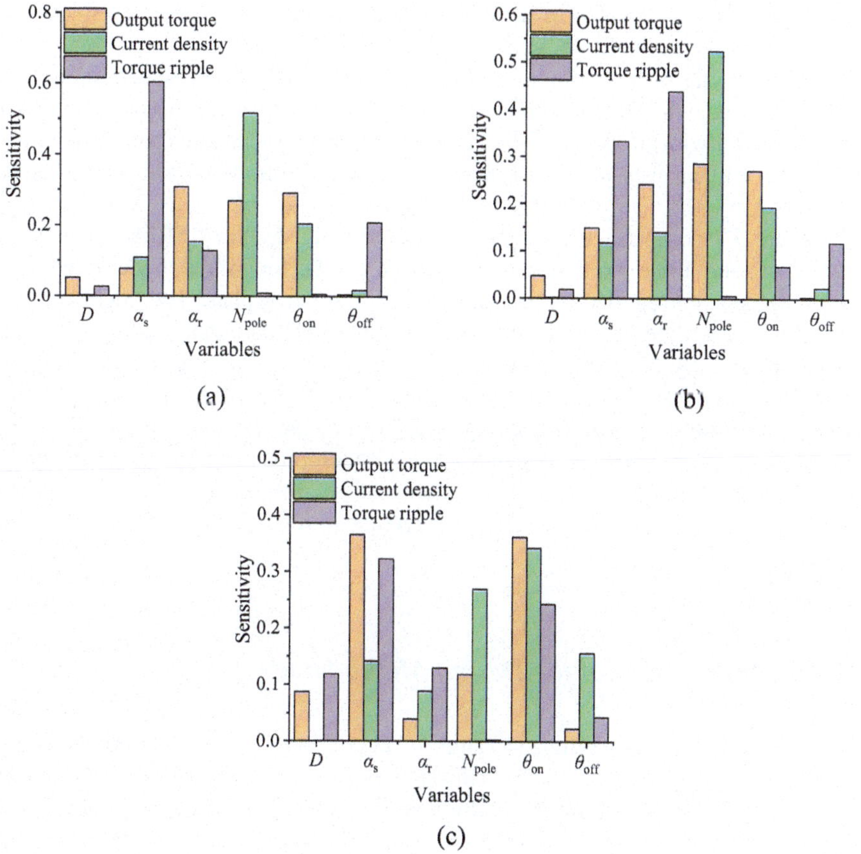

Fig. 3.4 Sensitivity indices for **a** 12-8 SRM, **b** 10-8 SRM, and **c** 18–20 SRM

Table 3.4 Comprehensive sensitivity results

Topology	D	α_s	α_r	N	θ_{on}	θ_{off}
12-8 SRM	0.03	0.24	0.21	0.26	0.18	0.07
10-8 SRM	0.02	0.19	0.27	0.27	0.19	0.04
18–20 SRM	0.07	0.29	0.08	0.13	0.32	0.07

The APC method is used to establish the finite element models in the optimization. The optimization models and intelligent algorithms are processed sequentially to achieve the Pareto front for the designed SRMs [20]. Figure 3.5 presents the optimization results of the three optimization objectives. All the optimization results presented in Fig. 3.5 form the Pareto optimal solutions, which are the optimal solutions of the multi-objective optimization. The designer can select a point from the Pareto solutions according to the specific design requirements. For better comparison and verification of the design methodology, the optimal solutions are selected from the Pareto front for each SRM with the same current density in this case study. The results are listed in Table 3.5, where *sf* is the slot fill factor.

The magnetic field distributions of the designed SRMs are presented in Fig. 3.6. As can be seen, when the stator magnetic poles are connected alternately in polarities NS, the motor with an odd-numbered phase will generate a long flux path. In contrast, the motor with an even-numbered phase will generate a short flux path with adjacent poles.

Figure 3.7 compares the torque of the optimally designed SRMs under rated operation, and Table 3.6 lists detailed data. It can be found that, under conditions with the same current density and slot fill factor, the five-phase 10/8 SRM exhibits the highest output torque, and the nine-phase 18–20 SRM outputs the lowest torque. The relationships between flux linkage and the current of each pole for the designed SRMs are presented in Fig. 3.8. Similar to the conclusion of the output torque, the five-phase 10/8 SRM exhibits the highest K_L value. The K_L values of each SRM are listed in Table 3.5, which correspond to the output torque performance. It should be noted that the actual K_L values listed in Table 3.5 are different from the preset values listed in Table 3.2. In theory, a smaller difference in K_L will lead to a more accurate result of the initial design.

3.4.3 *Verification of the Design Methodology*

According to the final optimized values of the designed SRMs, the output torque can be achieved based on (3.34). This step is a reverse verification of the design methodology by calculating the performance of the existing design variables. The comparison between the calculation values and the simulation values with the final optimal variables is tabulated in Table 3.6.

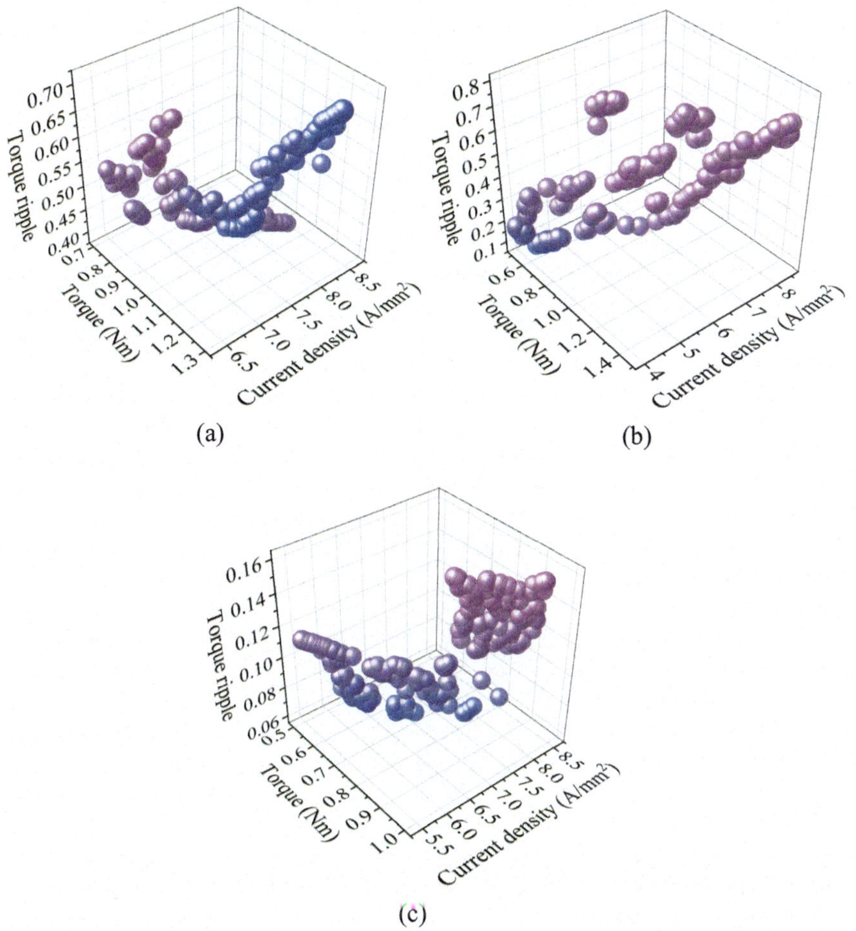

Fig. 3.5 Optimization results. **a** 12-8 SRM, **b** 10-8 SRM, and **c** 18–20 SRM

Table 3.5 The selected optimal values

Variable	12/8 SRM	10/8 SRM	18/20 SRM
α_s	0.49	0.44	0.36
α_r	0.33	0.33	0.31
N_{pole}	25	22	22
θ_{on} (°)	−0.8	−2.3	−1.4
T_{avg} (N m)	1.13	1.31	0.87
J_c (A/mm^2)	8	8	8
k_r (%)	57.57	58.96	11.79
K_L	5.5	5.6	2.6
sf	0.54	0.55	0.57

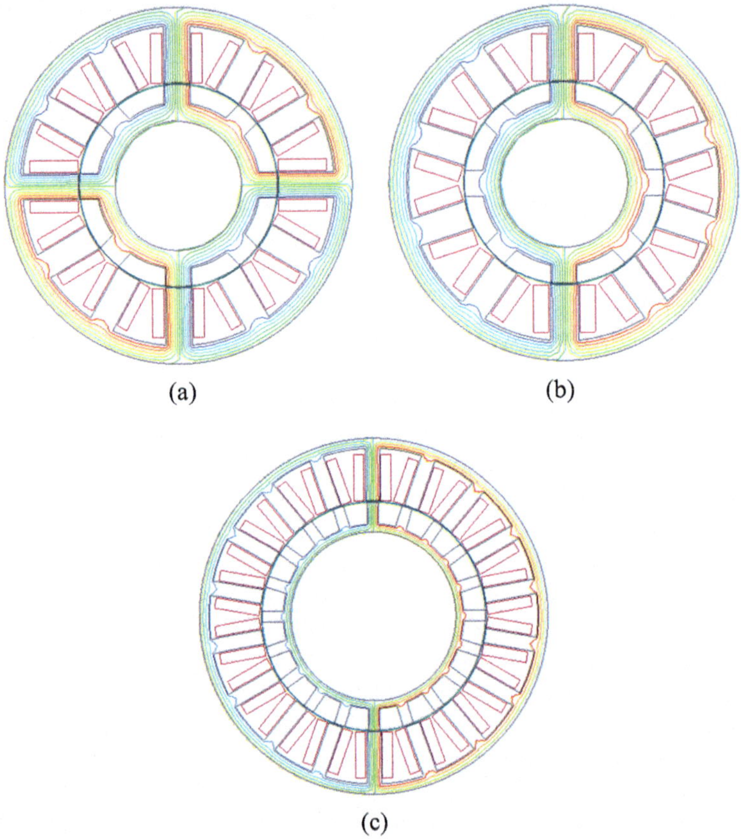

Fig. 3.6 Magnetic field distribution of the **a** 12/8 SRM, **b** 10/8 SRM, and **c** 18/20 SRM

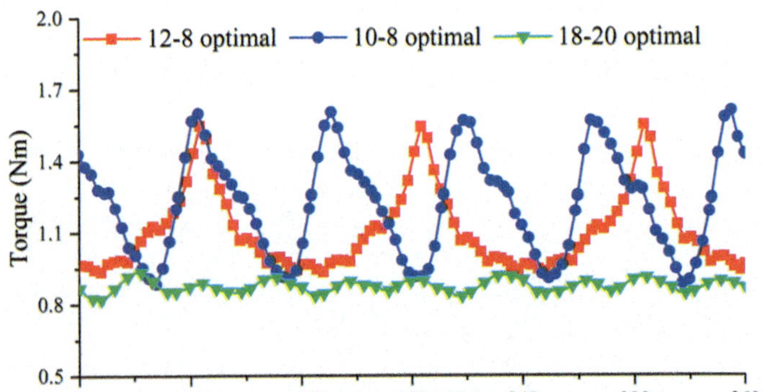

Fig. 3.7 Torque comparison between the three SRMs

Table 3.6 Comparison of calculated and optimal values

Topology	Torque T_{avg} (N m)	
	Calculation	Simulation
12/8 SRM	1.19	1.13
10/8 SRM	1.61	1.31
18/20 SRM	1.07	0.87

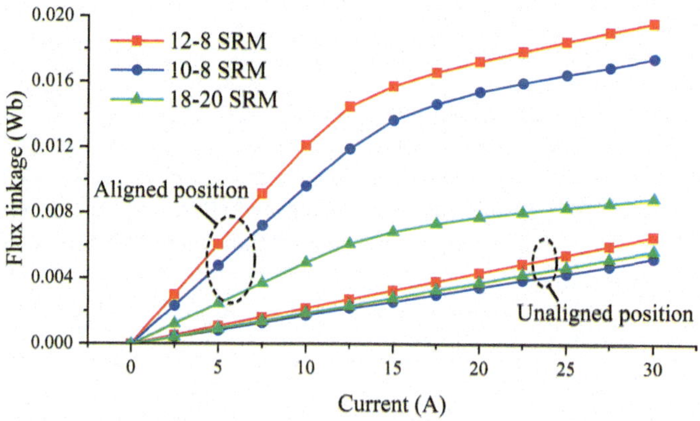

Fig. 3.8 The relationship between flux linkage and the current of each pole

The errors between the calculated and simulated values can be mainly explained by the following two reasons:

(1) The difference between the value of K_L.
In the initial design stage, we set the K_L for the three SRMs as 5, 4, and 3, respectively, while the actual values for the optimal solutions are 5.5, 5.6, and 2.6, respectively, as listed in Table 3.5. It can be found that the five-phase 10/8 SRM exhibits the maximum error between the preset and the actual values, which achieves 35%. Thus, this is one of the main reasons for the difference in calculated and simulated torques.

(2) Inconsistency in the conduction period between the initial and optimal designs.
In our design theory, we assume that the conduction angle is half of the period, while the actual value for the optimal design is less than this. This results in the calculation value being larger.

Considering the above two errors in the initial design stage, the errors between the calculated and the actual optimized values are accepted. Thus, the design methodology can be verified by the simulation results.

3.4.4 Experimental Validation

To verify the design optimization method, the three-, five-, and nine-phase SRM prototypes are manufactured, as shown in Fig. 3.9. The corresponding experiment platform is established, as shown in Fig. 3.10. The whole drive system is mounted on the test bed, which mainly includes the tested SRM, the torque transducer, and the hysteresis brake. The torque is measured by the torque transducer, and the load torque is provided by the hysteresis brake. The dSPACE 1401 is equipped to realize stable driving control, and the IGBT inverter is utilized to implement the turn-on and turn-off functions.

Figure 3.11 presents the operation characteristics of the three-, five-, and nine-phase SRMs in the APC mode under the rated condition. It can be found that, among the three SRMs, the five-phase SRM exhibits the maximum values of torque but also the largest torque ripple. Besides, with the increase of phase number, the torque ripple decreases. The output torques are consistent with the simulation results presented in Fig. 3.7, which verifies the design theory and the optimization method. The three-phase SRM exhibits the maximum current values, resulting in the terrible torque/current ratio among the three SRMs.

(a)

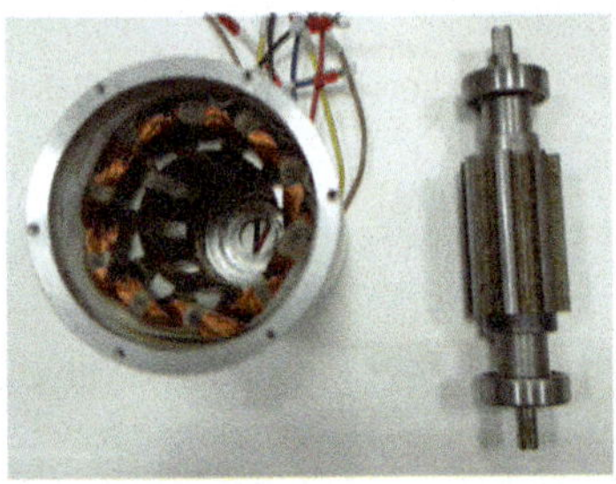
(b)

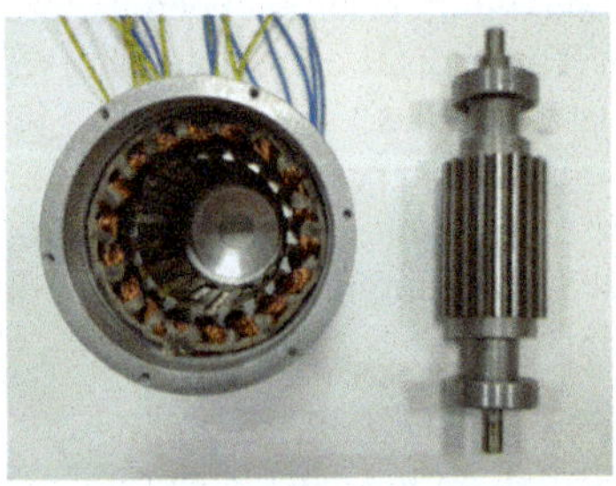
(c)

Fig. 3.9 The prototypes of stator and rotor. **a** 12/8 SRM, **b** 10/8 SRM, and **c** 18/20 SRM

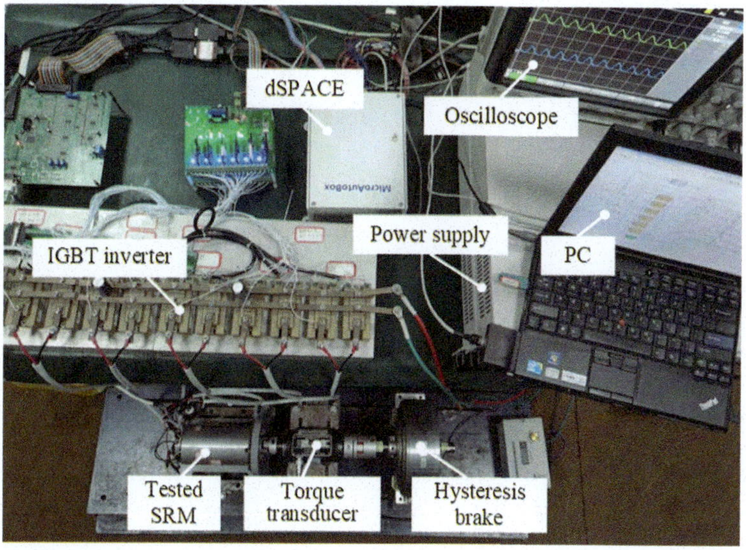

Fig. 3.10 Experimental test platform

3.5 Summary

This chapter presented design methods for SRMs, including the conventional design method and the proposed novel magnetic parameter design method. The proposed method was established based on the nonlinear characteristics of flux linkage for SRMs, which aims to establish an accurate calculation of the design parameters without overdependence on design experience. Three SRMs were initially designed according to the design method and further optimized by a multi-objective optimization method. Experimental results verified the effectiveness of the proposed method.

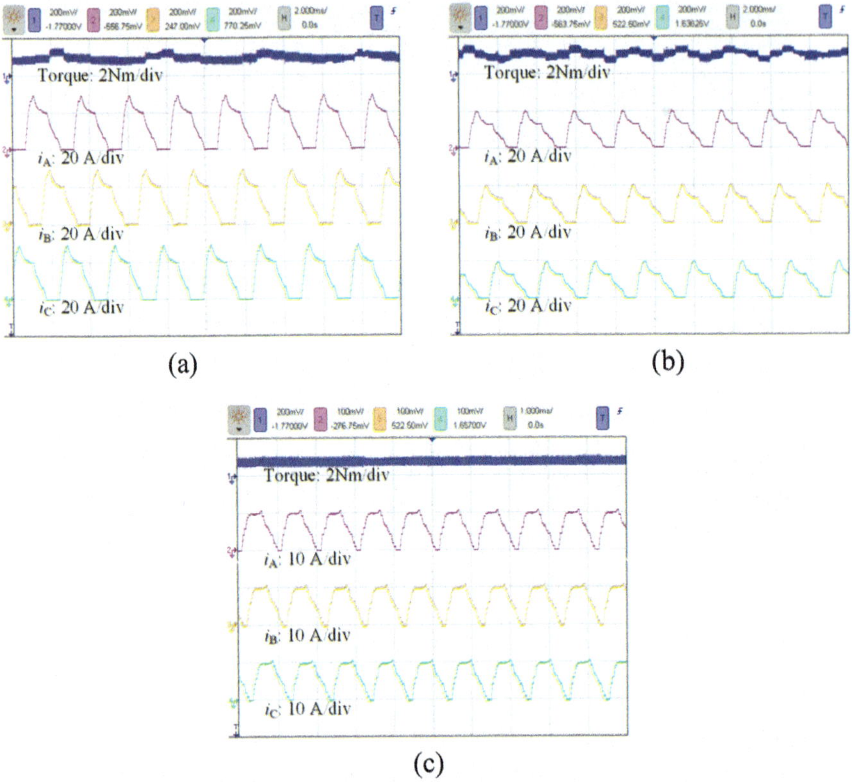

Fig. 3.11 Measured torque and current waveforms with APC mode at 24 V, 3000 rev/min. **a** Three-phase 12/8 SRM, **b** five-phase 10/8 SRM, and **c** nine-phase 18/20 SRM

References

1. Diao K, Sun X, Chen L, Cai Y, Wang H, Wu J. Direct torque control of a segmented switched reluctance motor for BSG in HEVs. In: 2019 3rd Conference on vehicle control and intelligence (CVCI). 2019. p. 1–6.
2. Diao K, Sun X, Lei G, Bramerdorfer G, Guo Y, Zhu J. System-level robust design optimization of a switched reluctance motor drive system considering multiple driving cycles. IEEE Trans Energy Convers. 2021;36(1):348–57. https://doi.org/10.1109/TEC.2020.3009408.
3. Li S, Zhang S, Habetler TG, Harley RG. Modeling, design optimization, and applications of switched reluctance machines—a review. IEEE Trans Ind Appl. 2019;55(3):2660–81. https://doi.org/10.1109/TIA.2019.2897965.
4. Song S, Xia Z, Fang G, Ma R, Liu W. Phase current reconstruction and control of three-phase switched reluctance machine with modular power converter using single DC-link current sensor. IEEE Trans Power Electron. 2018;33(10):8637–49. https://doi.org/10.1109/TPEL.2017.2782562.
5. Kim J, Kim R. Sensorless direct torque control using the inductance inflection point for a switched reluctance motor. IEEE Trans Ind Electron. 2018;65(12):9336–45. https://doi.org/10.1109/TIE.2018.2821632.

6. Sun X, Wang N, Cao Y, Guo D, Yao M, Sun Y, et al. Nonlinear fast modeling method of flux linkage and torque for a 128 switched reluctance motors. IEEE Trans Power Electron. 2024;39(5):5298–309. https://doi.org/10.1109/TPEL.2024.3363027.
7. Mihic DS, Terzic MV, Vukosavic SN. A new nonlinear analytical model of the SRM with included multiphase coupling. IEEE Trans Energy Convers. 2017;32(4):1322–34. https://doi.org/10.1109/TEC.2017.2707587.
8. Huang S, Cao G, He Z, Pan JF, Duan J, Qian Q. Nonlinear modeling of the inverse force function for the planar switched reluctance motor using sparse least squares support vector machines. IEEE Trans Ind Inform. 2015;11(3):591–600. https://doi.org/10.1109/TII.2015.2411438.
9. Hua H, Hua W. Analytical prediction of torque of switched reluctance machines considering nonlinear characteristics. IEEE Trans Ind Electron. 2021;1–1. https://doi.org/10.1109/TIE.2021.3051547
10. Diao K, Sun X, Bramerdorfer G, Cai Y, Lei G, Chen L. Design optimization of switched reluctance machines for performance and reliability enhancements: a review. Renew Sust Energy Rev. 2022;168. https://doi.org/10.1016/j.rser.2022.112785
11. Miller TJE. Optimal design of switched reluctance motors. IEEE Trans Ind Electron. 2002;49(1):15–27. https://doi.org/10.1109/41.982244.
12. Bilgin B, Emadi A, Krishnamurthy M. Design considerations for switched reluctance machines with a higher number of rotor poles. IEEE Trans Ind Electron. 2012;59(10):3745–56. https://doi.org/10.1109/TIE.2011.2141102.
13. Kurihara N, Bayless J, Sugimoto H, Chiba A. Noise reduction of switched reluctance motor with high number of poles by novel simplified current waveform at low speed and low torque region. IEEE Trans Ind Appl. 2016;52(4):3013–21. https://doi.org/10.1109/TIA.2016.2549993.
14. Sun X, Diao K, Lei G, Guo Y, Zhu J. Study on segmented-rotor switched reluctance motors with different rotor pole numbers for BSG system of hybrid electric vehicles. IEEE Trans Veh Technol. 2019;68(6):5537–47. https://doi.org/10.1109/TVT.2019.2913279.
15. Desai PC, Krishnamurthy M, Schofield N, Emadi A. Novel switched reluctance machine configuration with higher number of rotor poles than stator poles: concept to implementation. IEEE Trans Ind Electron. 2010;57(2):649–59. https://doi.org/10.1109/TIE.2009.2034678.
16. Schofield N, Long SA, Howe D, McClelland M. Design of a switched reluctance machine for extended speed operation. IEEE Trans Ind Appl. 2009;45(1):116–22. https://doi.org/10.1109/TIA.2008.2009506.
17. Qiao W, Diao K, Han S, Sun X. Design optimization of switched reluctance motors based on a novel magnetic parameter methodology. Electr Eng. 2022. https://doi.org/10.1007/s00202-022-01610-8.
18. Diao K, Sun X, Lei G, Guo Y, Zhu J. Multimode optimization of switched reluctance machines in hybrid electric vehicles. IEEE Trans Energy Convers. 2021;36(3):2217–26. https://doi.org/10.1109/tec.2020.3046721.
19. Diao K, Bramerdorfer G, Sun X, Yang Z, Han S. Multiobjective design optimization of a novel switched reluctance motor with unequal alternating stator yoke segments. IEEE Trans Transp Electrification. 2023;9(1):512–21. https://doi.org/10.1109/tte.2022.3174337.
20. Diao K, Sun X, Lei G, Guo Y, Zhu J. Multiobjective system level optimization method for switched reluctance motor drive systems using finite-element model. IEEE Trans Ind Electron. 2020;67(12):10055–64. https://doi.org/10.1109/TIE.2019.2962483.s.

Chapter 4
Control Methods for Switched Reluctance Motors

Abstract The nonlinear characteristics caused by the double salient structure limit the applications of switched reluctance motors (SRMs). Advanced control techniques have been developed to suppress torque ripples, enhance anti-disturbance ability, expand speed range, and improve efficiency. In this chapter, the benefits and shortcomings of control methods are classified and summarized. Besides, an improved direct torque control is presented.

Keywords Control methods · Direct torque control · Torque sharing function · Model predictive control

4.1 Introduction

Effective control of switched reluctance motors (SRMs) is paramount to harnessing their advantages and achieving optimal performance in various applications, such as industrial machinery, electric vehicles, and renewable energy systems. The unique characteristics of SRMs, such as their highly nonlinear behavior, variable magnetic saturation, and inherent torque ripples, present challenges that necessitate sophisticated control strategies to achieve smooth operation, high efficiency, and accurate speed and torque regulations [1–4].

Classical control methods for SRMs often involve open-loop control, where commutation angles are pre-determined based on operating conditions. These methods are relatively simple to implement but fall short in dynamically adapting to changes in load, speed, and other parameters. As a result, torque ripple and efficiency fluctuations may limit the motor's overall performance [5, 6].

Closed-loop control strategies come into play to tackle these limitations. The Proportional-Integral-Derivative (PID) controller can regulate speed and minimize torque ripple by adjusting the phase current profiles in response to feedback signals. However, the nonlinear characteristics of SRMs can challenge the effectiveness of conventional PID control, necessitating advanced control techniques [7].

X. Sun et al., *Multi-objective Design Optimization of Switched Reluctance Motor Drive Systems*, https://doi.org/10.1007/978-981-96-0672-6_4

Modern control approaches, such as torque sharing function (TSF) and direct torque control (DTC), enhance the adaptability of SRMs to varying conditions and provide improved performance in terms of torque precision, speed regulation, and energy efficiency [2, 8–10]. For instance, model predictive control (MPC) optimizes control inputs over a predictive horizon, considering constraints and objectives, thereby minimizing torque ripple and enhancing transient responses [11].

This section explores the diverse control strategies available for SRMs, encompassing conventional and advanced approaches. It serves as a guidepost for researchers, engineers, and practitioners seeking to harness the full potential of SRMs in various domains. The remaining sections are organized as follows. Section 4.2 presents and compares various control methods and further summarizes their features. Section 4.3 describes an example of an improved DTC, including the scheme, controller, and optimization algorithm, followed by the summary in Sect. 4.4.

4.2 Control Methods of Switched Reluctance Motors

This section reviews various control methods for SRMs, including their working principles, mathematical models, and implementation techniques. Besides, the advantages and limitations of each control method are discussed.

4.2.1 Conventional Control Strategies

(a) Chopping control

Two different methods, named current chopping control (CCC) and voltage pulse width modulation (PWM) control (VPWMC), are characterized by the object and process. CCC is realized by directly controlling the current, which is suitable for the low-speed range below the base speed. By controlling the switching devices, the current is limited between the given limits, while the turn-on and turn-off angles are constant. As for the VPWMC, the average value of the excitation voltage is changed by the effective time width of the control signal, modulating the phase voltage applied to the conducting phase. According to two hysteresis control techniques, it is possible to distinguish between the hard chopping and the soft chopping approaches. The soft chopper has fewer chopping times, less switching loss, and lower acoustic noise than the hard chopper. The torque is relatively stable due to the current featuring a flat-topped wave shape.

(b) Angular position control (APC)

APC is utilized for medium- and high-speed ranges. It adjusts the turn-on and turn-off angles under constant phase voltage. Optimal performance and efficiency can be achieved from tunable angles. However, when it is applied in the low-speed range, the peak current would exceed the limit, thereby easily leading to the burning of the power converter or the windings.

Taking advantage of the compensating features of CCC and APC, we can employ a combined control strategy to limit the peak current for a good start and speed regulation performance and effectively broaden the speed range of the drive system. However, the high torque ripple and the unsteady starting torque are the two main problems of this method.

4.2.2 Indirect Control

(a) Torque sharing function (TSF) control

The torque demand in TSF control is distributed among the various phases by combining hysteresis or PWM methods, aggregating individual phase contributions to the net torque. The resultant instantaneous torque tracks the expected torque by the position or closed-loop speed controller. The total torque synthesized is kept constant, thereby suppressing torque ripple. The block diagram of typical TSF control is shown in Fig. 4.1.

The control performance, including torque ripple and copper loss, is determined by the design of the TSF, and the general formula is described as [8]

$$\begin{cases} T_{k,\mathrm{ref}}(\theta) = T_{\mathrm{ref}} f_k(\theta), \; k = 1, 2, \ldots, m \\ \sum_{k=1}^{m} f_k(\theta) = 1, \qquad 0 \le f_k(\theta) \le 1 \end{cases} \tag{4.1}$$

where T_{ref} is the total reference torque, $T_{k,\mathrm{ref}}$, and f_k are the reference torque and distribution function of phase k, respectively, and m is the total number of phases. The four conventional expressions of TSF are linear, cosine, cubic, and exponential TSF. Taking the cosine type of TSF as an example, the typical waveform is shown in Fig. 4.2, where θ_{on}, θ_{off}, and θ_{ov} are the turn-on, turn-off, and overlap angles, respectively. The turn-on and overlap angles have been optimized based on the genetic

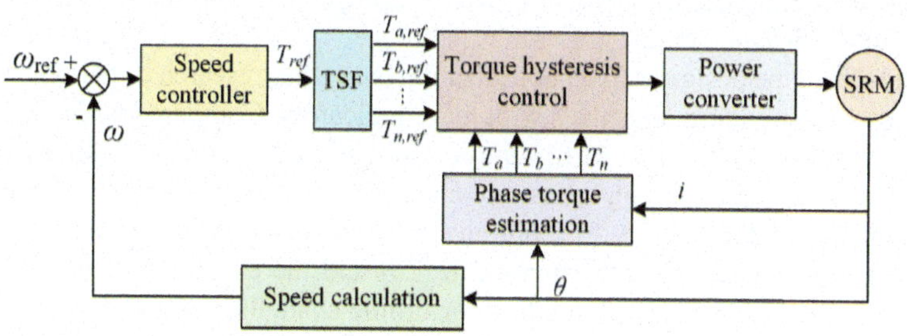

Fig. 4.1 The block diagram of TSF control

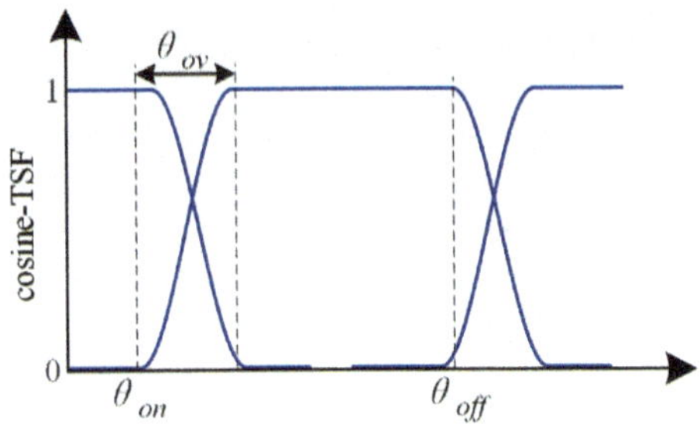

Fig. 4.2 The cosine type of TSF

algorithm [12]. TSF control strategy can effectively avoid large peak currents and inhibit excessive torque ripple. However, due to the slow current response, the torque distribution function cannot be tracked well during commutation. The reduced torque of the former phase cannot be compensated by the increased torque of the latter phase, and the constant total torque cannot be guaranteed. Torque ripples still have a particular effect under high speed and heavy load conditions.

Recent works regarding the TSF mainly focus on the torque ripple and copper loss reduction and the improvement of effective speed range through designing and optimizing the distribution function and optimization target. Related research on the design of distribution functions can be divided into analytical expression and dynamic allocation. Three analytical expressions were introduced in [13–15] to meet the minimal torque ripple and further optimization objectives, including a novel family of reference current waveforms with tunable parameters, a nonlinear TSF with the power function, and a nonlinear modulating factor. Similarly, a modified linear function and a non-unity piecewise quadratic TSF were established to reduce torque ripple in [16, 17].

The analytic formula TSF may still be unable to achieve complete tracking of the reference torque due to the limited rate of current change. Thus, the distribution function in the form of dynamic allocation was presented [18]. In [19–21], different strategies to compensate for current and torque errors during the commutation were presented. Analogously, a novel nonlinear TSF with demagnetization during commutation was proposed for smooth torque sharing between two adjacent phases [22]. Furthermore, several control algorithms, such as neural network control, fuzzy logic control, and iterative learning control, were employed to adjust the reference current and torque [23–25].

Aside from the design of TSF, optimization is another research direction based on defining corresponding objectives with more indicators. Compared with the consideration of phase current in [13, 26], a more sophisticated optimization objective with weight parameters, including the rate of change of current, was established in [27]. Different from TSFs in [13, 27], optimized TSFs were introduced through information on the phase voltage dynamics in [28] and the extension of the definition region in [29], respectively. A method for offline evaluation of objective functions was presented, leading to less computational time [30]. Table 4.1 lists the key aspects of each TSF approach to show the differences in the three categories.

Table 4.1 Comparison of TSF control

Approach	Reference	Speed range	Performance	Computational burden
Analytical expression	[12–17, 25, 38–40]	Low	Fewer metrics	Low
Dynamic allocation	[18–24]	High	Fewer metrics	Medium
Optimization	[13, 26–29]	High	Comprehensive	Heavy

(b) Current profiling technique

The current profiling technique is a control method that directly deals with the reference current, while TSF refers to reference torque as the control object. This method requires debugging to obtain the optimal current profile and does not rely on a lookup table. The phase current was set to rise and fall linearly in the commutation region [31]. Based on the initial base current profiles, the current profile was optimized and tuned with position and torque feedback using a high-bandwidth current regulator. An improved current profile with radial force sum flattening was presented in [22] to reduce acoustic noise and vibration. Aside from optimized current profiles, several hybrid measures were taken to suppress the torque ripple, such as the multiple reference frame controller, the commutation shift, and the rotor shape.

(c) Harmonic current injection

Including multiple additional harmonic current components to the reference current was proposed to suppress the torque ripple [32]. The current waveform can be specified by summarizing different harmonic currents, as shown in Fig. 4.3. The form of injected current I_{inj} can be given by

$$I_{inj} = \sum_{1}^{n} I_n \sin(nN\theta + P_n). \tag{4.2}$$

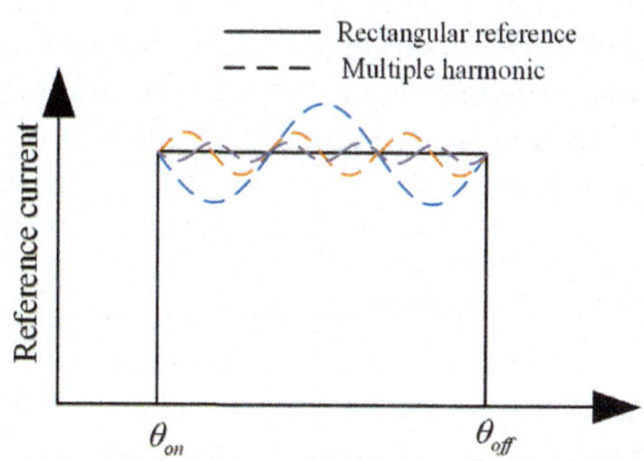

Fig. 4.3 Schematic representation of multiple harmonic injections

where I_n and P_n are the magnitudes and spatial phase of nth order injected current, respectively, and N is the lowest common multiple of stator and rotor pole numbers.

Related improvement studies have also been proposed to achieve satisfactory tracking. As reported in [33], the DC component mainly influenced the average torque and the first and second-order current harmonics. Furthermore, the fourth- and fifth-order current harmonics primarily affected the torque ripple. To reduce the torque ripple, several methods were developed for the broad application of SRMs, including a segmented method combined with control algorithms [34], a novel method modified into unipolar waveforms [35], and five different novel procedures based on optimizing the current profile [36]. To reduce acoustic noise and vibration, multi-objective optimization of the amplitude and phase of injection current was carried out [37].

4.2.3 Direct Control

Direct control with a torque feedback loop can be generally classified into direct instantaneous torque control (DITC) and DTC (direct torque control), which take the torque as the control objective.

(a) DITC

The typical block diagram of DITC consists of a speed controller, instantaneous torque calculation unit, torque hysteresis control unit, power converter, and SRM model, as shown in Fig. 4.4. The outer loop and the inner loop are the speed loop and the torque loop, respectively. The switch states of the conduction phase are determined according to the deviation of reference torque T_{ref} and instantaneous torque T within the preset rotor angle interval. DITC can obtain more accurate instantaneous torque, avoiding the complexity of indirect control [41]. Take phase B as an example. Figure 4.5 shows the conduction interval distribution, where any phase's working period is divided into regions I, II, and III according to the conduction angle. Figure 4.6 illustrates the switching tables of the single-phase and two-phase working conditions where S(k) is the switching signals of phase k, and $\Delta T_{\min}$ and $\Delta T_{\max}$ are inner hysteresis and outer hysteresis, respectively.

The DITC was first implemented on a four-quadrant SRM drive [42]. The PWM and predictive control are integrated into DITC for torque ripple reduction without offline-calculated current or flux profiles [43]. A torque-sharing algorithm was introduced to minimize ohmic losses [44] and acoustic noise [45]. Besides, several other technical approaches have been incorporated into DITC for performance improvements. For instance, as reported in [46, 47], a 4-level converter and a modular multi-level power converter were integrated with DITC to enhance dynamic performance and efficiency. Aside from converter innovation, a different switching strategy according to TSF [48], current control [49], parameter identification without rotor locking [50], a speed controller based on adaptive terminal sliding mode [51], and

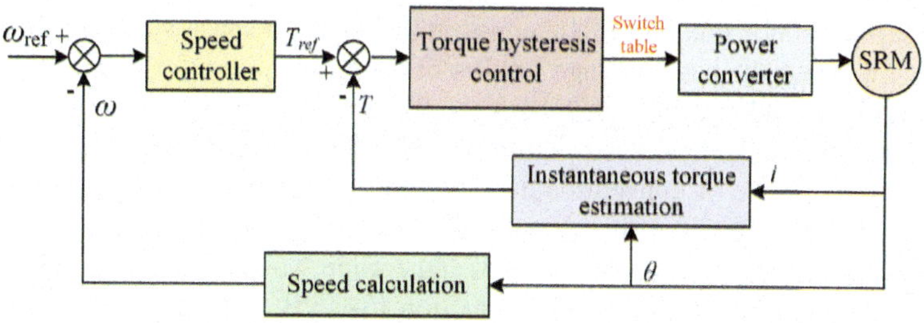

Fig. 4.4 The block diagram of DITC

Fig. 4.5 The distribution of conduction interval

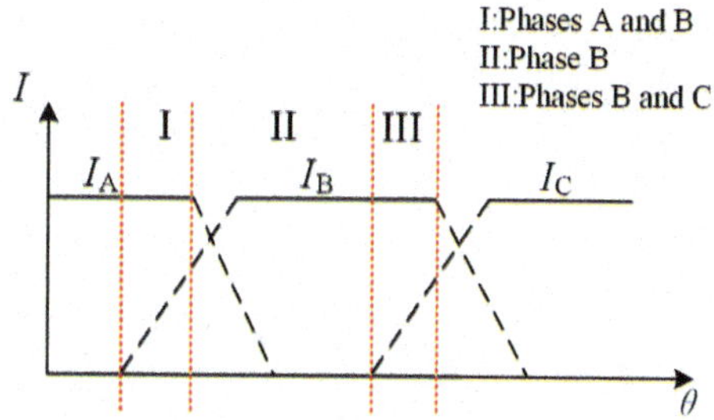

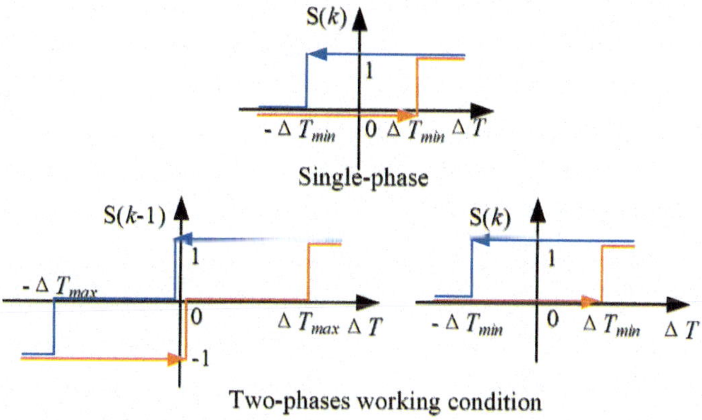

Fig. 4.6 The illustration of the switching tables

an adaptive dynamic commutation strategy [52] were utilized in DITC for broader speed range, lower torque ripple, and superior efficiency.

(b) DTC

The effect of DTC equipped with parallel torque and flux linkage control loop is more prominent. The block diagram of DTC control is shown in Fig. 4.7. The main idea is to keep the stator flux amplitude constant and select a space voltage vector [53]. Figure 4.8 shows the basic voltage vectors of the DTC system for a 12/8 SRM. The torque and flux linkage difference between the given value and the actual feedback value is limited to the hysteresis width, suppressing the torque ripple.

The performance of conventional DTC will deteriorate under high-speed operation conditions. To solve this drawback, a variable given flux corresponding to the relationship between the given flux and torque was designed for better operational efficiency and torque ripple reduction [54, 55]. Furthermore, an improved speed controller was developed in the optimized variable flux DTC to enhance the comprehensive performance within the speed range [56]. For voltage vector interval division, novel division ways and selection rules were established to suppress the torque ripple further, remove negative torque, and increase the torque-ampere ratio [10, 57, 58]. Especially in [10], the hysteresis control of flux linkage was eliminated, and a novel selection rule of voltage vectors was established, resulting in

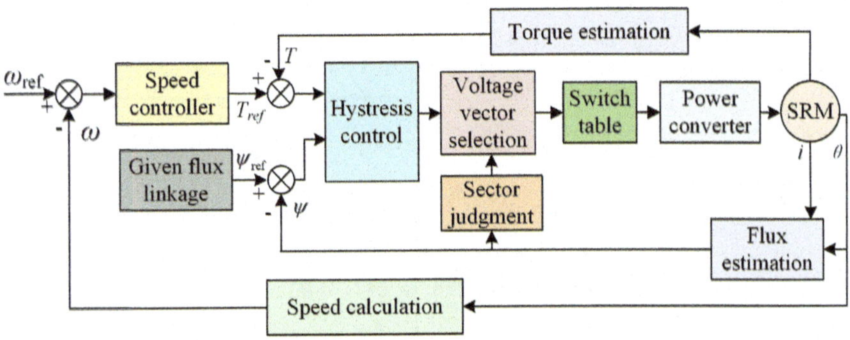

Fig. 4.7 The block diagram of DTC for SRMs

Fig. 4.8 Basic voltage vectors

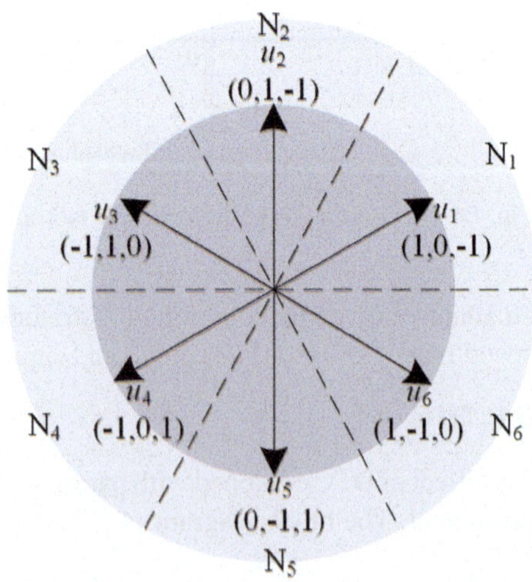

Table 4.2 Comparison of advanced DTC methods

References	Features	Advantages
[81, 82]	Variable given flux	Improve performance under high-speed operation
[83–85]	Improve voltage vector interval division	Suppress torque ripple, remove negative torque sequences
[83]	Elimination of flux loop	Reduce complexity
[86, 87]	Deadbeat direct control	Maintain the constant switching frequency, suppress torque ripple
[88–91]	Combination with other control techniques	Decrease the number of switching devices, suppress torque ripple

lower complexity. A deadbeat direct torque and flux-linkage control was employed to minimize the torque ripple and maintain a constant switching frequency [59, 60]. Similarly, novel converter [61], model predictive control [62] and sliding mode control [63] were introduced into DTC for better control performance.

Table 4.2 summarizes and compares the major improvements in DTC techniques for SRMs. The features and main advantages of each method are listed.

4.2.4 *Model Predictive Control*

The MPC uses a system model to predict the future output value over a limited future period based on the current state quantity. Then, an optimal control output can be determined by evaluating the cost function. The technique performs online correction and rolling optimization of the model according to feedback information.

MPC has been applied to control various electrical machines [64]. The application of MPC in SRMs is mainly predictive control of current and torque. It can be further subdivided into model predictive current control (MPCC) and model predictive torque control (MPTC). Figure 4.9 shows a typical block diagram of MPC.

The cost function can be defined as

$$J = \lambda_1 (x_{\mathrm{ref}}(k) - x(k+1))^2 + \lambda_2 (\Delta j)^2 \tag{4.3}$$

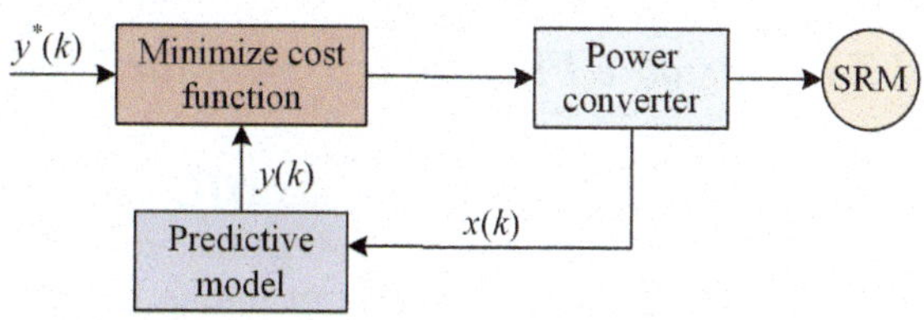

Fig. 4.9 The typical block diagram of MPC

where $x_{ref}(k)$ and $x_{ref}(k + 1)$ are reference values of the current and the next instant of time, respectively, λ_1 and λ_2 are the weighting coefficients of control objectives, and Δi is other control constraints, for example, copper loss. Different performance requirements can be included to define the cost function.

In terms of MPCC, a virtual-flux finite control set MPC was presented in [65], utilizing the flux-linkage tracking algorithm to control current indirectly. As for the cost function, a state graph for switching states limitation was introduced to reduce the number of commutations and computational burden. Different control methods have also been successfully applied to MPCC. For instance, in [66], an adaptive observer was incorporated in MPCC for a double stator SRM, thereby estimating and tracking the low-frequency variations of parameters like the inductance. Similarly, the inductance surface learning mechanism and inductance auto-calibration mechanism were combined with MPCC equipped with a Kalman filter estimator and compensation technique, which can cope with the measurement noise and uncertainties within the inductance profile. In addition to the above control techniques, the deadbeat control and current profiling techniques were applied in MPCC to obtain fixed switching frequency [67] and extended speed range [68], respectively. More related studies on MPCC can be found in [69].

MPTC was first employed for SRMs in [70], where the switches were directly manipulated, and the torque based on the nonlinear state model was regulated. Later, the virtual space vector modulation was reported in [71] for torque ripple suppression. To reduce the computational burden, a novel switching table along with the finite-state MPTC and an improved switching table based on inductance characteristics were established in [62, 72], respectively. In [73, 74], MPC and TSF were developed to reduce the torque ripple and avoid the frequency conversion caused by hysteresis control. Similarly, the predictive flux control and the MPTC were incorporated into DTC for desired performances like eliminating negative torque [75, 76]. Furthermore, low torque ripple, extended speed range, and four-quadrant operation can be achieved by combining online adaptive phase excitation, MPTC, and deadbeat predictive control [77, 78].

More detailed contributions about MPC can be found in [79, 80], which provide a systematic review. To show the difference between SRM control strategies, a list of several key aspects of each strategy is presented in Table 4.3.

4.3 Example of an Improved Direct Torque Control

4.3.1 Variable Flux DTC System for SRM

In the conventional structure of the DTC system, the torque can be controlled according to the change of phase relation between flux linkage and rotor position angle, thereby limiting the range of torque to reduce the torque. The stator flux amplitude is constant and related to the motor's operating performance. A subtle

Table 4.3 Comparison of control strategies

Control strategy	Typical method	Speed range	Torque ripple	Complexity	Switching angles
Conventional control	Chopping control	Low/Medium	High	Simple	Not required
	APC	High	High	Simple	Required
Indirect torque control	TSF	Low/Medium	Low/Medium	Medium	Required
	Current profiling	High	Low/Medium	Complex	Not required
	Harmonic current injection	Medium	Medium	Medium	Required
Direct torque control	DITC	High	Low	Simple	Required
	DTC	Low/Medium	Medium	Medium	Not required
Model predictive control	MPC	Medium	Low	Complex	Not required

difference in the amplitude can cause a relatively significant change in the relevant parameters, such as current and torque. Given that the value of the flux amplitude is set within a certain range, the system can thus achieve the goal of speed regulation. The value beyond the range will lead to high torque ripples, causing oscillations or non-convergence of the speed response.

The instantaneous torque of SRM can be described as

$$T \approx i\frac{\partial \psi(i,\theta)}{\partial \theta} \tag{4.4}$$

where T, i, θ, and Ψ are the instantaneous torque, current, rotor position angle, and flux linkage, respectively.

Equation (4.4) shows that a relatively large value of flux amplitude is a prerequisite for low-speed operation. In contrast, a relatively small value is required in high-speed stable operation to prevent high current amplitude, high torque ripple, and flux over-saturation. The optimal flux amplitude in the conventional DTC varies with the reference speed. Besides, the flux amplitude gradually decreases with the speed increase based on the tuning experience. Thus, the influence of flux variation on the performance is considered for torque ripple suppression.

Since the essence of the speed regulation system is torque control and the core of DTC is flux linkage, the variation of flux linkage is consistent with torque. Based on the above analysis, the reference torque is adjusted by the proportional coefficient as the control signal of flux linkage so that the flux linkage can be automatically adjusted according to the torque variation demand. The relationship between the reference flux linkage Ψ^* and T_e^* is approximatively set as $\Psi^* = k_1\, T_e^* + k_2$, where k_1 and

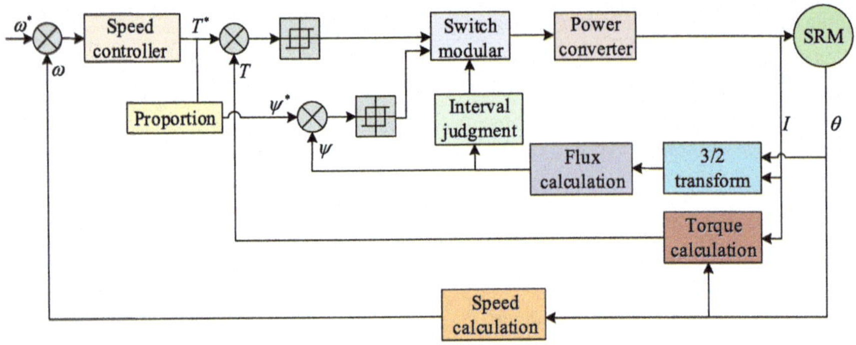

Fig. 4.10 A block diagram of the VF-DTC

k_2 are tunable coefficients. Figure 4.10 illustrates a block diagram of the VF-DTC system. The outer ring of the control system is the speed loop, and the inner ring is the parallel flux linkage loop and torque loop.

4.3.2 *Design of Speed Controller*

Since too many parameters need to be tuned in nonlinear ADRC, linear ADRC (LADRC) is introduced to solve the problems of engineering application and parameter tuning, which is simpler and easier to implement.

The mechanical movement equation of the SRM model can be described as

$$T_e^* - T_L - F \cdot \omega - J\frac{d\omega}{dt} = 0 \tag{4.5}$$

where J represents the moment of inertia, ω is the actual running speed, T_e* and T_L are the reference torque and the load torque, respectively, and F represents the frictional coefficient.

Equation (4.5) shows that the speed controller can be designed as a first-order LADRC based on the first-order mathematical model. The LADRC consists of linear tracking differentiator (LTD), linear extended state observer (LESO), and linear state error feedback (LSEF). To ease the contradiction between the given discontinuous signal and feedback continuous signal, LTD is chosen as the first-order inertia element $1/(1 + \tau s)$.

The LESO, which is the key component of LADRC, can estimate the disturbance in the system. According to (4.5), we have

$$\dot{\omega}_r = \frac{\mathrm{d}\omega_r}{\mathrm{d}t} = \frac{T_e^* - F \cdot \omega - T_L}{J}. \tag{4.6}$$

Furthermore, (4.6) can be written as

$$\dot{\omega}_r = f_r + b_0 T_e^* \tag{4.7}$$

where b_0 is the tunable coefficient, f_r represents the disturbance and can be expressed as

$$f_r = \frac{T_e^* - F \cdot \omega_r - T_L}{J} - b_0 T_e^*. \tag{4.8}$$

Selecting $x_1 = \omega_r$, $x_2 = f_r$, and $u = T_e^*$, the state equation is given by

$$\dot{x}_1 = x_2 + b_0 u. \tag{4.9}$$

In matrix form, (4.9) can be expressed as

$$\dot{x} = Ax + Bu + Ed \tag{4.10}$$

where $d = \dot{f}_r$, $x = \begin{bmatrix} x_1 \\ x_2 \end{bmatrix}$, $A = \begin{bmatrix} 0 & 1 \\ 0 & 0 \end{bmatrix}$, $B = \begin{bmatrix} b_0 \\ 0 \end{bmatrix}$, $E = \begin{bmatrix} 0 \\ 1 \end{bmatrix}$.

Combining the design theory of state observer (4.10), we have

$$\dot{z} = Az + Bu + L(x_1 - z_1) \tag{4.11}$$

where x_1 is the actual value of speed ω_r, z_1 the estimate of speed $\hat{\omega}_r$, and L the gain matrix, and $L = [\beta_1, \beta_2]^T$.

Equation (4.11) can be further expressed as

$$\begin{cases} \dot{\hat{\omega}}_r = \hat{f}_r + b_0 T_e^* + \beta_1(\omega_r - \hat{\omega}_r) \\ \dot{\hat{f}}_r = \beta_2(\omega_r - \hat{\omega}_r) \end{cases} \tag{4.12}$$

where $\hat{f}_r$ is an estimation of the disturbance.

According to (4.12), the expression of the state equation can be derived as

$$\begin{cases} e_1 = z_1 - x_1 \\ \dot{z}_1 = z_2 - \beta_1 e_1 + b_0 T_e^* \\ \dot{z}_2 = -\beta_2 e_1 \end{cases} \tag{4.13}$$

where $z_2 = \hat{f}_r$, and e_1 is the estimated error.

The formula indicates that the disturbance estimation is obtained by the time integral of the estimated error. The utilization ratio of the estimated error is low during observation of disturbance, which leads to a slow observation speed of disturbance. Thus, the proportional gain can be added to improve the estimated error's observation speed and utilization rate. The state equation of the improved LESO can be designed

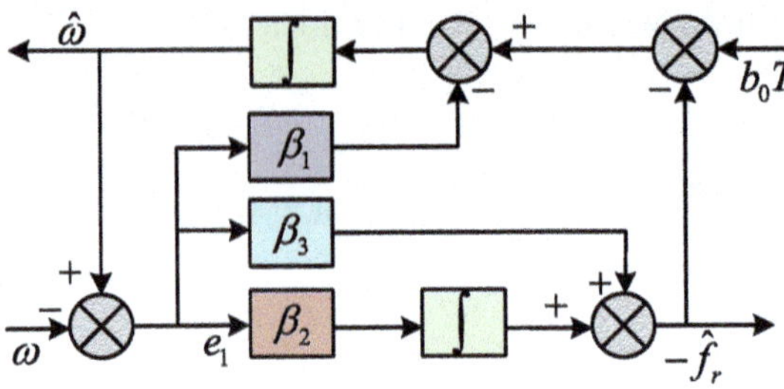

Fig. 4.11 A schematic diagram of the improved LESO

as

$$\begin{cases} e_1 = z_1 - x_1 \\ \dot{z}_1 = z_2 - \beta_1 e_1 + b_0 T_e^* \\ \dot{z}_2 = -\beta_2 e_1 - \beta_3 \dot{e}_1 \end{cases} \tag{4.14}$$

where β_3 is the differential gain coefficient of the estimated error.

The structure diagram of improved LESO is shown in Fig. 4.11.

Based on the improved LESO, gain matrix L can be modified as $L = [\beta_1 + \beta_3, \beta_2]^T$.

Subtract (4.11) from (4.10), we get

$$\dot{e} = A_e e + Ed \tag{4.15}$$

where $A_e = \begin{bmatrix} -\beta_1 - \beta_3 & 1 \\ -\beta_2 & 0 \end{bmatrix}$. Then, the characteristic polynomial of A_e is

$$\lambda(s) = s^2 + (\beta_1 + \beta_3)s + \beta_2. \tag{4.16}$$

According to the Routh-Hurwitz stability criterion, the system stability can be guaranteed when $\beta_1 > 0, \beta_2 > 0, \beta_3 > 0$.

LSEF is introduced to counteract disturbances and then $b_0 T_e^* = u_0 - \hat{f}_r$ is selected. When the estimated error of disturbance is ignored, (4.7) can be simplified as

$$\dot{\omega}_r = f_r + u_0 - \hat{f}_r \approx u_0 \tag{4.17}$$

where u_0 is the linear control rate output.

After disturbance compensation, the speed loop can be equivalent to an integral part. Select the linear control law as the proportional control,

$$u_0 = k(\omega_r^* - \hat{\omega}_r) \tag{4.18}$$

where ω_r^* represents reference speed, and k is the proportionality coefficient.

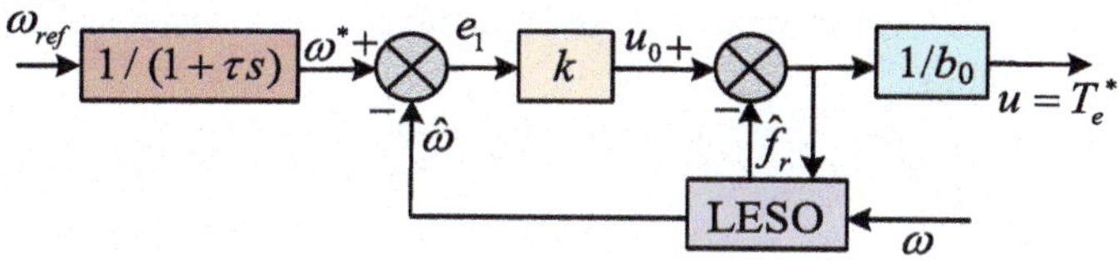

Fig. 4.12 A block diagram of the improved LADRC

Figure 4.12 shows the block diagram of the speed loop using improved LADRC. The method based on the bandwidth is adopted for tuning parameters as follows

$$\begin{cases} k = \omega_{rcv} \\ \tau = 1/\omega_{rcv} \\ \beta_3 = \beta_1 = 2\omega_{r0} \\ \beta_2 = \omega_{r0}^2 \end{cases}. \tag{4.19}$$

where ω_{rcv} and ω_{r0} represent the bandwidth of the speed loop and LESO, which satisfies $\omega_{r0} = 5{-}10\omega_{rcv}$.

4.3.3 *Hybrid Optimization Algorithm for Autotuning*

(a) Gray wolf optimization algorithm

The gray wolf optimization algorithm (GWOA) mimics the strict hierarchy and predatory behavior of gray wolves in nature, which has been widely used in some specific applications [30]. In the social hierarchy, gray wolves are divided into the wolf α, β, δ, and θ from high to low, which represent the best solution, the second-best solution, the third-best solution, and the candidate solution, respectively. The mathematical model of the encircling behavior of gray wolf can be expressed as

$$\vec{D} = \left|\vec{C} \cdot \vec{X}_p - \vec{X}\right| \tag{4.20}$$

$$\vec{X}(t+1) = \left|\vec{X}_p - \vec{A} \cdot \vec{D}\right| \tag{4.21}$$

where t is the current iteration number, $\vec{D}$ is the distance between the gray wolf and prey, $\vec{X}$ and $\vec{X}_p$ are the position vectors of the gray wolf and prey.

The coefficient vectors, $\vec{C}$ and $\vec{A}$, can be calculated as

$$\begin{cases} \vec{A} = 2 \cdot \vec{a} \cdot \vec{r}_1 - \vec{a} \\ \vec{C} = 2 \cdot \vec{r}_2 \end{cases} \tag{4.22}$$

where $\vec{r}_1$ and $\vec{r}_2$ are random vectors in [0, 1], and $\vec{a}$ is linearly decreased from 2 to 0 throughout iterations.

The mathematical model of the hunting behavior of gray wolves is shown as

$$\vec{D}_p = \left| \vec{C} \cdot \vec{X}_p - \vec{X} \right| \tag{4.23}$$

$$\vec{X}_n = \vec{X}_p - \vec{A} \cdot \vec{D}_p \tag{4.24}$$

$$\vec{X}(t+1) = \frac{\vec{X}_1 + \vec{X}_2 + \vec{X}_3}{3} \tag{4.25}$$

where the subscript p denotes α, β, and δ, respectively, $n = 1$, 2 and 3, and $\vec{X}_n$ is the movement instructions given by α, β, and γ, respectively.

Compared with other algorithms, the GWOA stands out due to its strong local search capability and high convergence rate. However, GWOA finds it hard to achieve satisfactory results in solving complex optimization problems, likely falling into a local solution.

(b) Improved coyote optimization algorithm

The coyote optimization algorithm (COA) mimics the coyote's social life, growth, birth, and death, which focuses on the social structure and experiences exchange instead of social hierarchy and dominance rules. The COA performs better in handling complex optimization problems with several groups and global searching capability. However, the COA may not work well in terms of efficiency, stability, and computation complexity. The improved COA (ICOA) is designed to improve global searching capability, convergence rate, and operability. The three main steps of ICOA are listed below.

The first step is to initialize the packs N_p, coyotes N_c, parameter ranges, iterations, and the total number of coyotes N where $N = N_p \times N_c$. Those initial social conditions are set randomly for each coyote. In this paper, parameters N_p and N_c are adjusted dynamically to achieve better operability. Then, Divide the population of coyotes into groups. The groups are assigned randomly after dynamic parameter adjustment, and each coyote's social condition is calculated. Based on dynamic adjustment, the process of coyotes being driven out and accepted by the group can be eliminated.

The second step is the growth of coyotes in each pack, which is influenced by the best coyote *alpha*, the cultural tendency *cult*, and two coyotes r_1 and r_2 selected at random. The correlation expression can be described as

$$cult_j = \begin{cases} O_{(N_c+1)/2,j}, & N_c \text{ is odd} \\ (O_{N_c/2,j} + O_{N_c/2+1,j})/2, & \text{otherwise} \end{cases} \tag{4.26}$$

where j is the dimension in the range [1, D], $cult_j$ is the cultural tendency, and O_j is the ranked social conditions of all coyotes.

The growing method of COA can be expressed as

$$new\_soc_c = soc_c + d_1 \cdot \delta_1 + d_2 \cdot \delta_2 \tag{4.27}$$

$$\delta_1 = alpha - soc_{r1} \tag{4.28}$$

$$\delta_2 = cult_j - soc_{r2} \tag{4.29}$$

where soc_{r1} and soc_{r2} are two random coyotes' social conditions in a pack; d_1 and d_2 are two evenly distributed random numbers inside [0, 1].

The Gaussian numbers are adopted instead of the uniform random numbers in the ICOA to increase the search range and improve global searching capacity. Thus, (4.27) can be further designed as

$$new\_soc_c = soc_c + rn_1 \cdot \delta_1 + rn_2 \cdot \delta_2 \tag{4.30}$$

where rn_1 and rn_2 are the Gaussian numbers.

The third step is the birth and death of the newborn coyotes. The newborn coyotes are influenced by the genetic and environmental factors of coyotes from randomly selected parents, which can be described as

$$newborn_j = \begin{cases} soc_{r1,j}, \; r_j < P_s \; or \; j = j_1 \\ soc_{r2,j}, \; r_j \geq P_s + P_a \; or \; j = j_2 \\ R_j, \quad\quad otherwise \end{cases} \tag{4.31}$$

$$P_s = 1/D \tag{4.32}$$

$$P_a = (1 - P_s)/2 \tag{4.33}$$

where j_1 and j_2 are two random dimensions, and R_j is a random number inside the variable range. P_s and P_a are the scatter and association probabilities, determining the heredity and variation of the newborn coyotes, respectively.

(c) Hybrid Optimization Algorithm

Based on the respective characteristics of GWOA and ICOA, a hybrid optimization algorithm (HOA) combining the two search forms is designed to compensate for their disadvantages and achieve superior performance. The flowchart of HOA is shown in Fig. 4.13.

In early iterations, the two updating methods generate new solutions in an equal probability way to ensure diversity and exploration capability. As for the latter iterations, one method is utilized to strengthen the local exploring capability. The above analysis reveals that GWOA behaves well in convergence performance while ICOA exhibits better performance in terms of searching capability. Choosing the

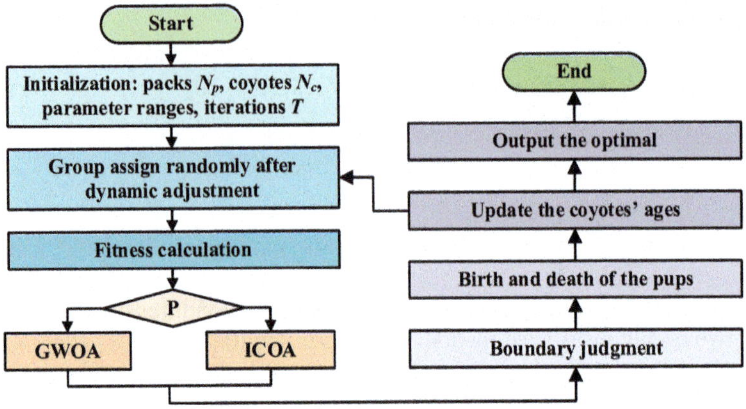

Fig. 4.13 The flowchart of HOA

GWOA in early iterations and ICOA in later iterations can improve the optimization performance due to their complementary advantages. The probability P is selected as

$$\mathrm{P} = \begin{cases} 0.55 & T < \text{iteration\_ max}/2 \\ 0.45 & T > \text{iteration\_ max}/2 \end{cases}. \tag{4.34}$$

As for the parameters, including packs N_p and coyotes N_c, it should be ensured that both the optimization's time and accuracy are considered. The HOA parameters are selected in Table 4.4 to balance the growth effect and global searching capability. The coefficients k1 and k2 are set inside $[1 \times 10^{-6}, 1]$ to find the optimal solution in an extensive range. Besides, the parameters of the speed controller, including ω_{rcv}, ω_{r0}, and b_0, are selected within $[1 \times 10^{-3}, 1 \times 10^{6}]$.

The fitness function should be defined to select the tunable parameters automatically with HOA in the whole control system. The major objective is to achieve a satisfactory dynamic of speed and torque in various conditions. The minimization of the sampling speed error and torque error will guarantee superior performance. Thus, the designed function should be compatible with the control objectives as follows

Table 4.4 The HOA parameters

Parameters	Values
Packs, N_p	10
Coyotes, N_c	10
Coefficients, k_1, k_2	$[1 \times 10^{-6}, 1]$
Parameters, ω_{rcv}, ω_{r0}, b_0	$[1 \times 10^{-3}, 1 \times 10^{6}]$

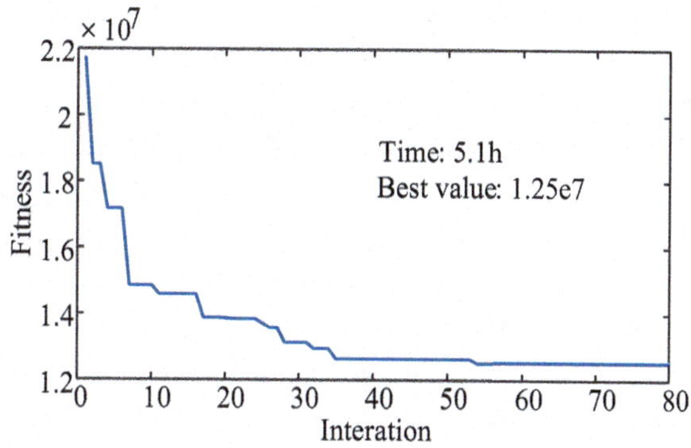

Fig. 4.14 A convergence process of fitness value of HOA

$$F = \frac{1}{N}\sum_{n=0}^{N}\left[|\omega_r^*(n) - \omega_r(n)|nTs + |T_e^*(n) - T_e(n)|nTs\right] \tag{4.35}$$

where ω_r^* is the reference speed, T_e represents the actual feedback torque, and n and T_s are the sampling number and the sampling period, respectively.

Figure 4.14 shows a convergence process of the fitness value of the HOA. The maximum number of iterations is 80, and the best fitness index is $F = 1.25 \times 10^7$ for the VF-DTC system.

4.3.4 Simulation Results

To clearly show the speed control performance, anti-interference ability, robustness, and torque ripple suppression of the proposed VF-DTC using HOA, simulations with conventional DTC system with PI (method 1), VF-DTC with LADRC (method 2), VF-DTC with PI using HOA (method 3), and VF-DTC with LADRC using HOA (proposed method) are compared. Besides, the research mainly focuses on solving the problem of performance degradation of conventional DTC in high-speed operation range. Thus, various operating conditions in the medium to high-speed range are selected. The performance of different control systems will be validated by analyzing the speed and torque waveforms.

The DTC system of the SRM drive simulation platform is built utilizing Matlab/Simulink. The reference speed is set as 3000 rev/min. The parameters of the simulation model and the specifications of SRM are listed in Table 4.5.

Figure 4.15 shows the specific iteration processes of the speed and torque with constant load torque at rated speed in the 5th, 20th, 30th, and 80th iterations. It can be seen that the best performance is exhibited in the 80th iteration in terms of dynamic response and torque ripple.

Table 4.5 Specifications of the SRM

Parameters	Values
Number of phases	3
Stator/Rotor poles	12/8
Rated speed (r/min)	3000
Load torque T_L (N)	0.8
DC-link voltage (V)	48
Sampling period (s)	1e−5

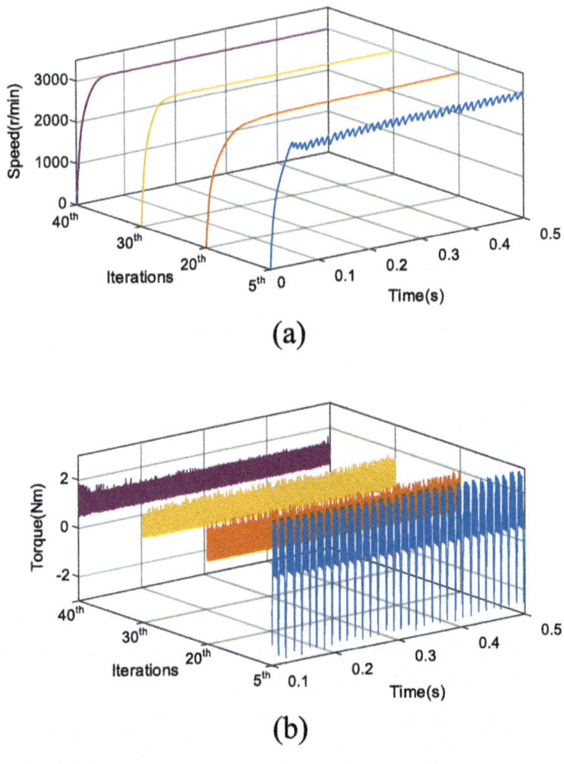

Fig. 4.15 Evolution in the 5th, 20th, 30th, and 80th iterations. **a** Speed, and **b** torque

As for the parameters of method 1, the proportional parameter P and the integral parameter I are 0.0007 and 0.016, respectively. The parameters k_1, k_2, ω_{cv}, ω_0, and b_0 of method 2 are set as 0.006, 0.01, 3096, 27,400, and 28,500, and those of the proposed method are 0.0063, 0.00935, 3617.14, 22,794.26, and 27,257.28, respectively. The parameters k_1, k_2, P, and I of method 3 are 0.0061, 0.012, 0.00068, and 0.0164. The parameters of methods 1 and 2 are selected using the dichotomy method. Concretely, the initial debugging scope is determined based on the stability principle. Then, the dichotomy method is used for parameter debugging. Finally, the parameters that improve dynamic performance in speed and torque are selected.

Case 1: Condition under constant load torque and rated speed

Figure 4.16 shows the simulation results of speed and torque using four control methods in this condition. As shown, three VF-DTC methods can track the reference speed, while conventional DTC with constant flux can only reach 2500 rev/min, instead of 3000 rev/min. In addition, the torque ripple range ΔT of DTC is larger than VF-DTC, which reflects the torque ripple suppression ability. As for the three VF-DTC methods, it can be seen that the settling times are 0.13 s, 0.09 s, and 0.07 s, respectively. The torque ripple ranges ΔT are 1.14 N m, 0.97 N m, and 0.8 N m, respectively. As shown, the performance of the proposed method behaves well in acceleration and torque ripple suppression.

The flux linkage vector locus under 0.8 N m at rated speed is shown in Fig. 4.17. It is found that the flux linkages of three VF-DTC methods are variable, while the given reference flux linkage of DTC is 0.033 Wb.

Case 2: Condition under medium-speed change

Figure 4.18 shows the simulation results when the speed changes from 1300 to 1500 rev/min, then to 1400 rev/min with a constant load torque (0.8 N m). The dynamic response time of the four methods is 0.18 s, 0.06 s, 0.12 s, and 0.03 s, respectively. The response time of method 1 is significantly higher than that of the other three methods when speed changes. Meanwhile, the speed has no significant effect on the torque ripple when using VF-DTC, while the torque ripple increases with the speed increase when using DTC. It can be seen that the proposed method exhibits the best performance in terms of dynamic response and the lowest torque ripples in the medium speed range.

Case 3: Condition under high-speed change

The high-speed change condition is considered for the simulation with a constant load torque (0.8 N m). Since the DTC cannot track the high speed well, no more simulations and experiments about this condition are compared. Figure 4.19 illustrates the dynamic responses of three VF-DTC methods. In the simulation, there are two changes in the speed (from 2500 to 3000 rev/min at 0.2 s and from 3000 to 2400 rev/min at 0.35 s). As shown, three VF-DTC methods take about 0.05 s to reach the steady state when changes happen. The torque has significant oscillations when the speed varies. Furthermore, the proposed method can remarkably reduce the torque ripples compared to the other methods.

Case 4: Condition under load torque change at medium-speed

To further compare the anti-disturbance performance of the three control methods, the simulation results of speed and torque in load torque change condition (from 0.8 to 1.2 N) at medium-speed are shown in Fig. 4.20. In this condition, the overshooting of the speed waveform pulsation using the proposed method is 51 r/min, which is reduced significantly compared with the other methods. It can be seen that the speed overshoots of methods 1 and 3 using the PI controller are obvious. Moreover, the response time of methods 1 and 3 takes about 0.2 s to return to the normal speed,

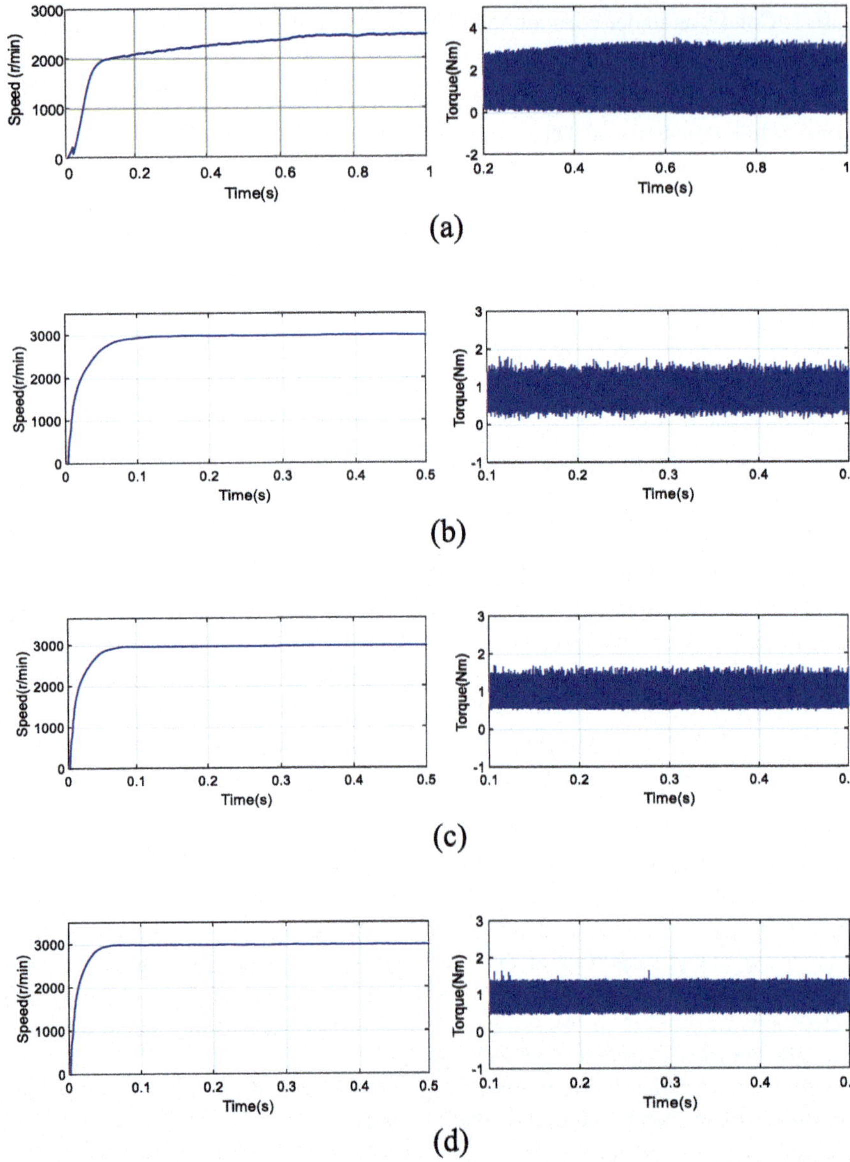

Fig. 4.16 Speed and torque response under rated conditions of four control methods: **a** method 1, **b** method 2, **c** method 3, **d** the proposed method

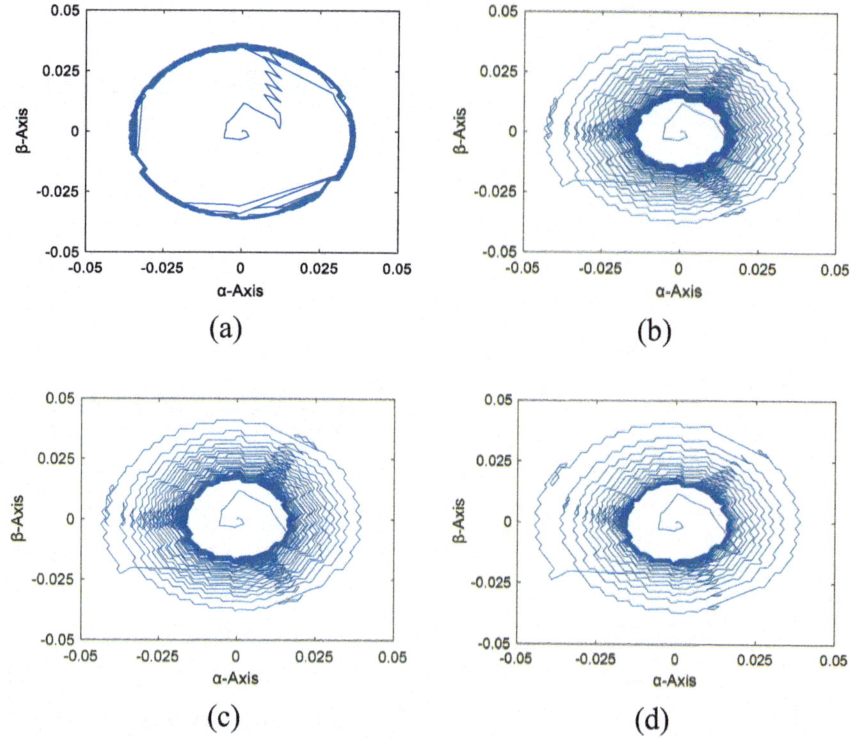

Fig. 4.17 Flux linkage vector locus under 0.8 N m at rated speed: **a** method 1, **b** method 2, **c** method 3, **d** the proposed method

which is lower than the other two methods. Furthermore, the torque ripples under different control methods decrease with the increase in load torque. Therefore, it reflects that VF-DTC using HOA has a better disturbance rejection capability than VF-DTC without HOA, and the anti-disturbance capability of methods 1 and 3 with the PI controller is unsatisfactory. Figure 4.21 shows the phase current of the proposed method when the load torques are 0.8 and 1.2 N m.

Finally, to clearly show the performance comparison of the four control methods, Table 4.6 tabulates the main performance results of the simulation for the four operational conditions. In summary, the proposed method can provide much better control performance in response time, robustness, anti-disturbance capability, and torque ripple reduction than the others.

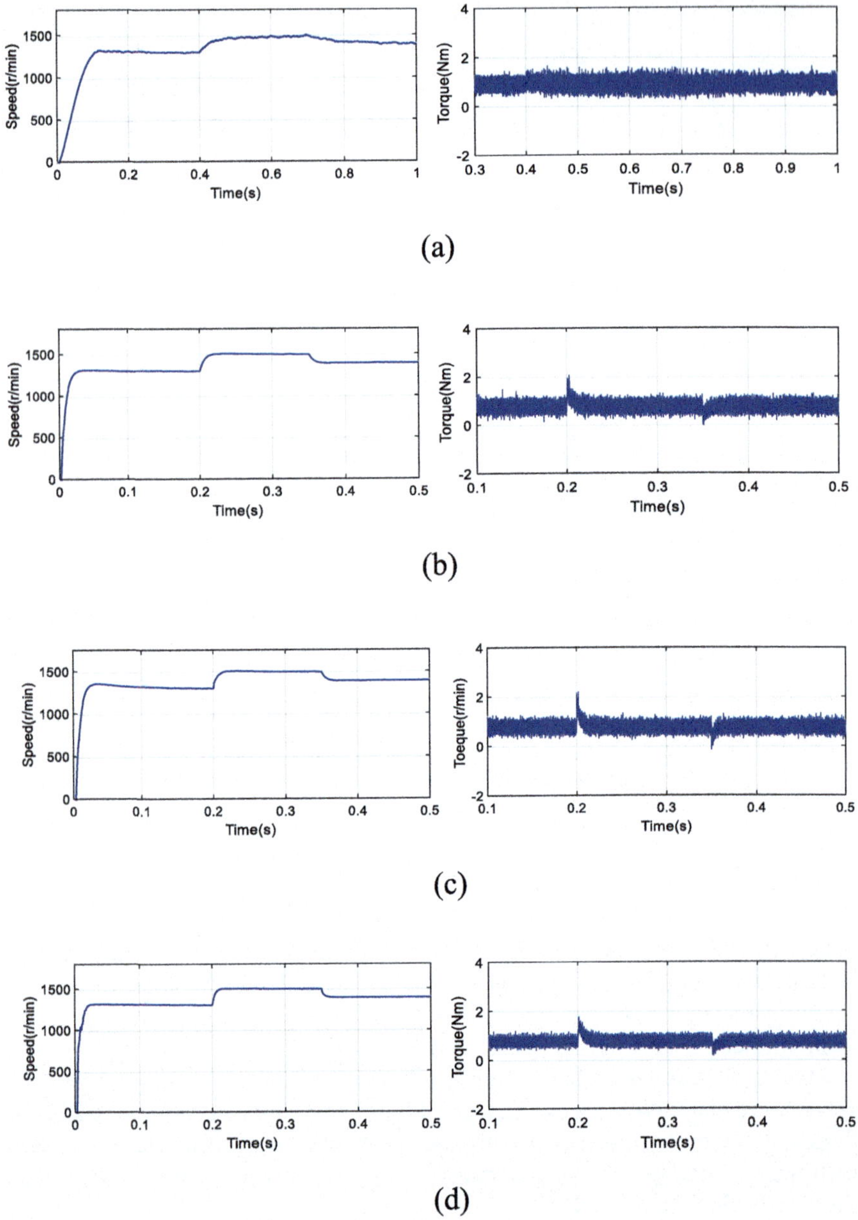

Fig. 4.18 Speed and torque response when the speed varies in the medium-speed range: **a** method 1, **b** method 2, **c** method 3, **d** the proposed method

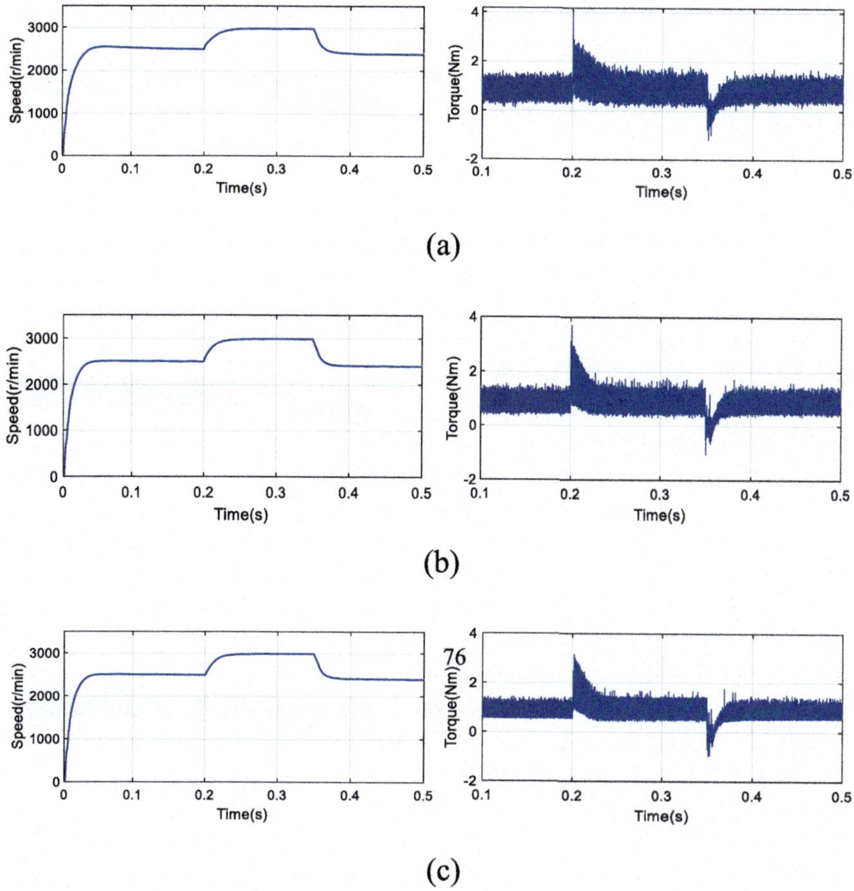

Fig. 4.19 Speed and torque response when the speed varies in the high-speed range: **a** method 2, **b** method 3, **c** proposed method

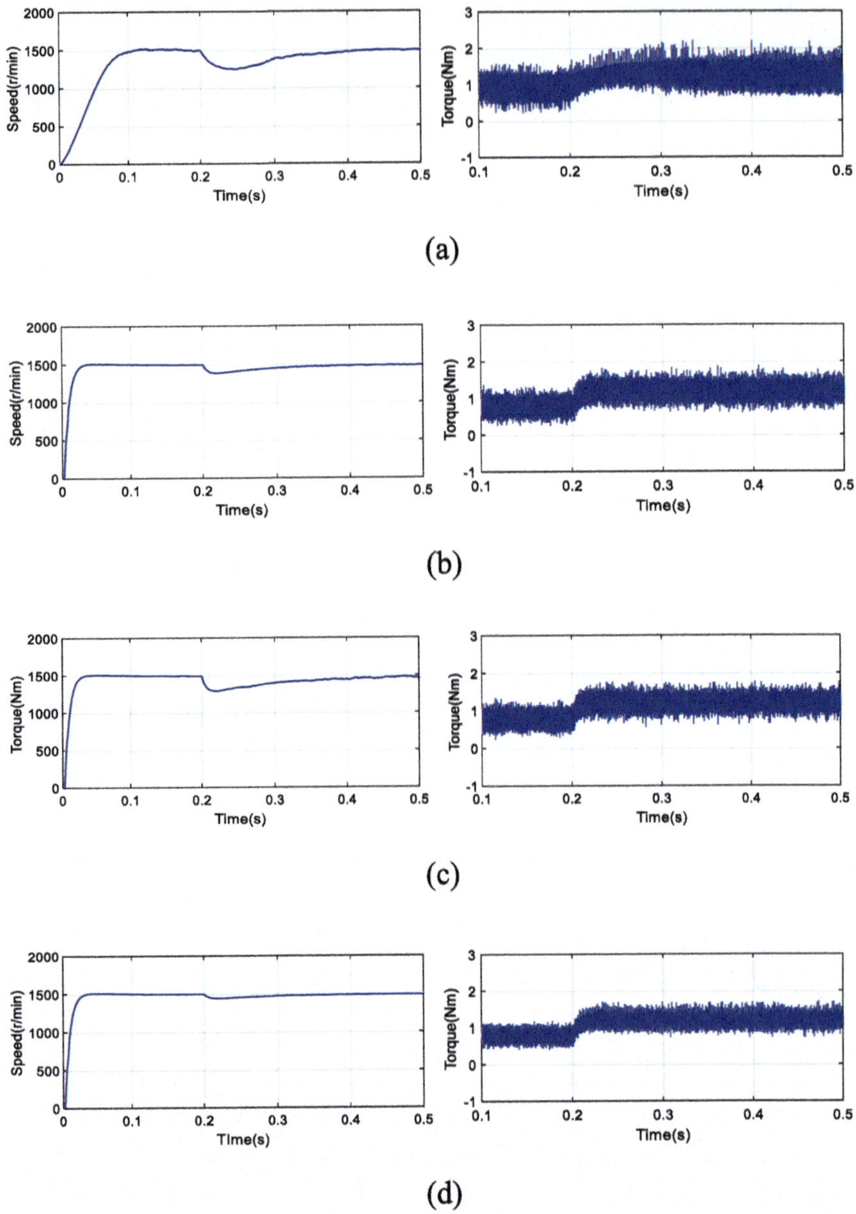

Fig. 4.20 Speed and torque response under load torque change at medium speed: **a** method 1, **b** method 2, **c** method 3, and **d** the proposed method

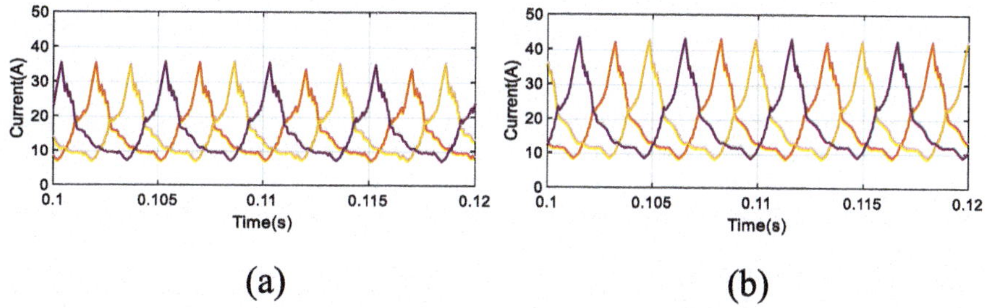

Fig. 4.21 Phase current response using the proposed method with different load torques. **a** 0.8 N m, and **b** 1.2 N m

Table 4.6 Performance results of the simulation

Case	Parameters	Method 1	Method 2	Method 3	Proposed method
Case 1	Setting time (s)	–	0.13	0.09	0.07
	Torque range (N m)	–	1.14	0.97	0.8
Case 2	Setting time (s)	0.18	0.06	0.12	0.03
	Speed response time (s)	0.16	0.04	0.04	0.03
Case 3	Speed response time (s)	–	0.06	0.05	0.05
Case 4	Response time (s)	0.21	0.17	0.2	0.12
	Overshoot (r/min)	250	107	243	51

4.4 Summary

This chapter summarizes the control methods for SRMs. Each method's features, advantages, and drawbacks were tabulated and compared. An improved DTC is investigated to find the optimal performances, in which the variable DTC system, the speed controller design, and the hybrid optimization algorithm are presented.

References

1. Ge L, Fan Z, Du N, Huang J, Xiao D, Song S. Model predictive torque and force control for switched reluctance machines based on online optimal sharing function. IEEE Trans Power Electron. 2023;38(10):12359–64. https://doi.org/10.1109/TPEL.2023.3295577.
2. Feng L, Sun X, Guo D, Yao M, Diao K. Advanced torque sharing function strategy with sliding mode control for switched reluctance motors. IEEE Trans Transp Electrif. 2024;10(1):2302–11. https://doi.org/10.1109/TTE.2023.3287159.
3. Xiao D, Ye J, Fang G, Xia Z, Emadi A. Magnetic-characteristic-free high-speed position-sensorless control of switched reluctance motor drives with quadrature flux estimators. IEEE J Emerg Sel Top Power Electron. 2022;10(1):220–35. https://doi.org/10.1109/jestpe.2021.3088064.

4. Diao K, Sun X, Lei G, Bramerdorfer G, Guo Y, Zhu J. System-level robust design optimization of a switched reluctance motor drive system considering multiple driving cycles. IEEE Trans Energy Convers. 2021;36(1):348–57. https://doi.org/10.1109/tec.2020.3009408.
5. Feng L, Sun X, Bramerdorfer G, Zhu Z, Cai Y, Diao K, et al. A review on control techniques of switched reluctance motors for performance improvement. Renew Sust Energ Rev. 2024;199:114454. https://doi.org/10.1016/j.rser.2024.114454.
6. Diao K, Sun X, Bramerdorfer G, Cai Y, Lei G, Chen L. Design optimization of switched reluctance machines for performance and reliability enhancements: a review. Renew Sust Energy Rev. 2022;168. https://doi.org/10.1016/j.rser.2022.112785
7. Viola J, Angel L, Sebastian JM. Design and robust performance evaluation of a fractional order PID controller applied to a DC motor. IEEE/CAA J Autom Sin. 2017;4(2):304–14. https://doi.org/10.1109/JAS.2017.7510535.
8. Feng L, Sun X, Yang Z, Diao K. Optimal torque sharing function control for switched reluctance motors based on active disturbance rejection controller. IEEE/ASME Trans Mechatron. 2023;1–9. https://doi.org/10.1109/TMECH.2023.3240986
9. Sun X, Feng L, Zhu Z, Lei G, Diao K, Guo Y, et al. Optimal design of terminal sliding mode controller for direct torque control of SRMs. IEEE Trans Transp Electrification. 2022;8(1):1445–53. https://doi.org/10.1109/tte.2021.3111889.
10. Yan N, Cao X, Deng Z. Direct torque control for switched reluctance motor to obtain high torque-ampere ratio. IEEE Trans Ind Electron. 2019;66(7):5144–52. https://doi.org/10.1109/tie.2018.2870355.
11. Li W, Cui Z, Ding S, Chen F, Guo Y. Model predictive direct torque control of switched reluctance motors for low-speed operation. IEEE Trans Energy Convers. 2022;37(2):1406–15. https://doi.org/10.1109/tec.2021.3131870.
12. Xue XD, Cheng KWE, Ho SL. Optimization and evaluation of torque-sharing functions for torque ripple minimization in switched reluctance motor drives. IEEE Trans Power Electron. 2009;24(9):2076–90. https://doi.org/10.1109/TPEL.2009.2019581.
13. Vujičić VP. Minimization of torque ripple and copper losses in switched reluctance drive. IEEE Trans Power Electron. 2012;27(1):388–99. https://doi.org/10.1109/tpel.2011.2158447.
14. Li C, Zhang C, Liu J, Bian D. A high-performance indirect torque control strategy for switched reluctance motor drives. Math Probl Eng. 2021;2021. https://doi.org/10.1155/2021/6618539
15. Rana AK, Teja AVR. A mathematical torque ripple minimization technique based on a nonlinear modulating factor for switched reluctance motor drives. IEEE Trans Ind Electron. 2022;69(2):1356–66. https://doi.org/10.1109/TIE.2021.3063871.
16. Mousavi-Aghdam SR, Moradi A, Dolatkhah AM. Torque ripple reduction of Switched reluctance motor using improved torque sharing functions. In: 2017 Iranian conference on electrical engineering (ICEE). 2017. p. 1043–7.
17. Park K, Liu X, Chen Z. A non-unity torque sharing function for torque ripple minimization of switched reluctance generators. In: 2013 15th European conference on power electronics and applications (EPE). 2013. p. 1–10.
18. Tseng KJ, Shuyu C. A SRM variable speed drive with torque ripple minimization control. In: APEC 2001 Sixteenth annual IEEE applied power electronics conference and exposition (Cat No01CH37181), vol. 2. 2001. p. 1083–9.
19. Ye J, Bilgin B, Emadi A. An extended-speed low-ripple torque control of switched reluctance motor drives. IEEE Trans Power Electron. 2015;30(3):1457–70. https://doi.org/10.1109/tpel.2014.2316272.
20. Sun Q, Wu J, Gan C, Hu Y, Si J. OCTSF for torque ripple minimisation in SRMs. IET Power Electron. 2016;9(14):2741–50. https://doi.org/10.1049/iet-pel.2016.0270.
21. Lee DH, So-Yeon A, Ahn JW, Jang-Mok K. Modified TSF for the high speed switched reluctance motor. In: 2011 IEEE International symposium on industrial electronics. 2011. p. 655–60.
22. Lee D-H, Liang J, Lee Z-G, Ahn J-W. A simple nonlinear logical torque sharing function for low-torque ripple SR drive. IEEE Trans Ind Electron. 2009;56(8):3021–8. https://doi.org/10.1109/TIE.2009.2024661.

23. Sahoo SK, Panda SK, Jian-Xin X. Indirect torque control of switched reluctance motors using iterative learning control. IEEE Trans Power Electron. 2005;20(1):200–8. https://doi.org/10.1109/TPEL.2004.839807.
24. Lin Z, Reay DS, Williams BW, He X. Torque ripple reduction in switched reluctance motor drives using B-spline neural networks. IEEE Trans Ind Appl. 2006;42(6):1445–53. https://doi.org/10.1109/TIA.2006.882671.
25. Ro H-S, Lee K-G, Lee J-S, Jeong H-G, Lee K-B. Torque ripple minimization scheme using torque sharing function based fuzzy logic control for a switched reluctance motor. J Electr Eng Technol. 2015;10(1):118–27. https://doi.org/10.5370/JEET.2015.10.1.118.
26. Xia Z, Bilgin B, Nalakath S, Emadi A. A new torque sharing function method for switched reluctance machines with lower current tracking error. IEEE Trans Ind Electron. 2021;68(11):10612–22. https://doi.org/10.1109/tie.2020.3037987.
27. Ye J, Bilgin B, Emadi A. An offline torque sharing function for torque ripple reduction in switched reluctance motor drives. IEEE Trans Energy Convers. 2015;30(2):726–35. https://doi.org/10.1109/tec.2014.2383991.
28. Changhwan C, Seungho K, Yongdae K, Kyihwan P. A new torque control method of a switched reluctance motor using a torque-sharing function. IEEE Trans Magn. 2002;38(5):3288–90. https://doi.org/10.1109/TMAG.2002.802295.
29. Li H, Bilgin B, Emadi A. An improved torque sharing function for torque ripple reduction in switched reluctance machines. IEEE Trans Power Electron. 2019;34(2):1635–44. https://doi.org/10.1109/tpel.2018.2835773.
30. Bober P, Ferkova Z. Comparison of an off-line optimized firing angle modulation and torque sharing functions for switched reluctance motor control. Energies. 2020;13(10). https://doi.org/10.3390/en13102435
31. Wallace RS, Taylor DG. A balanced commutator for switched reluctance motors to reduce torque ripple. IEEE Trans Power Electron. 1992;7(4):617–26. https://doi.org/10.1109/63.163641.
32. Stephenson JM, Hughes A, Mann R. Torque ripple minimisation in a switched reluctance motor by optimum harmonic current injection. IEE Proc Elect Power Appl. 2001;148(4):322–8. https://doi.org/10.1049/ip-epa:20010480.
33. Zhu ZQ, Lee B, Huang L, Chu W. Contribution of current harmonics to average torque and torque ripple in switched reluctance machines. IEEE Trans Magn. 2017;53(3):1–9. https://doi.org/10.1109/TMAG.2016.2633477.
34. Ma M, Ling F, Li F, Liu F. Torque ripple suppression of switched reluctance motor by segmented harmonic currents injection based on adaptive fuzzy logic control. IET Electr Power Appl. 2020;14(2):325–35. https://doi.org/10.1049/iet-epa.2019.0027.
35. Huang L, Zhu ZQ, Feng J, Guo S, Li Y, Shi JX. Novel current profile of switched reluctance machines for torque density enhancement in low-speed applications. IEEE Trans Ind Electron. 2020;67(11):9623–34. https://doi.org/10.1109/TIE.2019.2952801.
36. Shaked NT, Rabinovici R. New procedures for minimizing the torque ripple in switched reluctance motors by optimizing the phase-current profile. IEEE Trans Magn. 2005;41(3):1184–92. https://doi.org/10.1109/TMAG.2004.843311.
37. Gundogmus O, Sozer Y, Vadamodala L, Kutz J, Tylenda J, Wright RL. Current harmonics injection method for simultaneous torque and radial force ripple mitigation to reduce acoustic noise and vibration in SRMs. In: 2019 IEEE Energy conversion congress and exposition (ECCE). 2019. p. 7091–7.
38. Ilić-Spong M, Miller TJE, MacMinn SR, Thorp JS. Instantaneous torque control of electric motor drives. In: 1985 IEEE Power electronics specialists conference. 1985. p. 42–8.
39. Schramm DS, Williams BW, Green TC. Torque ripple reduction of switched reluctance motors by phase current optimal profiling. In: PESC'92 Record 23rd annual IEEE power electronics specialists conference, vol. 2. 1992. p. 857–60.
40. Husain I, Ehsani M. Torque ripple minimization in switched reluctance motor drives by PWM current control. IEEE Trans Power Electron. 1996;11(1):83–8. https://doi.org/10.1109/63.484420.

41. Inderka RB, Doncker RWAAD. DITC-direct instantaneous torque control of switched reluctance drives. IEEE Trans Ind Appl. 2003;39(4):1046–51. https://doi.org/10.1109/TIA.2003.814578
42. Fuengwarodsakul NH, Menne M, Inderka RB, Doncker RWD. High-dynamic four-quadrant switched reluctance drive based on DITC. IEEE Trans Ind Appl. 2005;41(5):1232–42. https://doi.org/10.1109/TIA.2005.853381.
43. Neuhaus CR, Fuengwarodsakul NH, Doncker RWD. Predictive PWM-based direct instantaneous torque control of switched reluctance drives. In: 2006 37th IEEE Power electronics specialists conference. 2006. p. 1–7.
44. Brauer HJ, Hennen MD, Doncker RWD. Control for polyphase switched reluctance machines to minimize torque ripple and decrease ohmic machine losses. IEEE Trans Power Electron. 2012;27(1):370–8. https://doi.org/10.1109/TPEL.2011.2161332.
45. Klein-Hessling A, Hofmann A, De Doncker RW. Direct instantaneous torque and force control: a control approach for switched reluctance machines. IET Electr Power Appl. 2017;11(5):935–43. https://doi.org/10.1049/iet-epa.2016.0515.
46. Liang J, Lee DH, Ahn JW. Direct instantaneous torque control of switched reluctance machines using 4-level converters. IET Electr Power Appl. 2009;3(4):313–23. https://doi.org/10.1049/iet-epa.2008.0002.
47. Song S, Peng C, Guo Z, Ma R, Liu W. Direct instantaneous torque control of switched reluctance machine based on modular multi-level power converter. In: 2019 22nd International conference on electrical machines and systems (ICEMS). 2019. p. 1–6.
48. Zeng H, Chen H, Shi J. Direct instantaneous torque control with wide operating range for switched reluctance motors. IET Electr Power Appl. 2015;9(9):578–85. https://doi.org/10.1049/iet-epa.2015.0087.
49. Husain T, Elrayyah A, Sozer Y, Husain I. Unified control for switched reluctance motors for wide speed operation. IEEE Trans Ind Electron. 2019;66(5):3401–11. https://doi.org/10.1109/TIE.2018.2849993.
50. Yao S, Zhang W. A simple strategy for parameters identification of SRM direct instantaneous torque control. IEEE Trans Power Electron. 2018;33(4):3622–30. https://doi.org/10.1109/TPEL.2017.2710137.
51. Sun X, Feng L, Diao K, Yang Z. An improved direct instantaneous torque control based on adaptive terminal sliding mode for a segmented-rotor SRM. IEEE Trans Ind Electron. 2021;68(11):10569–79. https://doi.org/10.1109/TIE.2020.3029463.
52. Sun Q, Wu J, Gan C. Optimized direct instantaneous torque control for SRMs with efficiency improvement. IEEE Trans Ind Electron. 2021;68(3):2072–82. https://doi.org/10.1109/TIE.2020.2975481.
53. Cheok AD, Fukuda Y. A new torque and flux control method for switched reluctance motor drives. IEEE Trans Power Electron. 2002;17(4):543–57. https://doi.org/10.1109/TPEL.2002.800968.
54. Guo H. Considerations of direct torque control for switched reluctance motors. In: 2006 IEEE International symposium on industrial electronics. 2006. p. 2321–5.
55. Zhao X, Xu A, Zhang W. Research on DTC system with variable flux for switched reluctance motor. CES Trans Elect Mach Syst. 2017;1(2):199–206. https://doi.org/10.23919/TEMS.2017.7961342.
56. Feng L, Sun X, Tian X, Diao K. Direct torque control with variable flux for an SRM based on hybrid optimization algorithm. IEEE Trans Power Electron. 2022;37(6):6688–97. https://doi.org/10.1109/TPEL.2022.3145873.
57. Shah V, Alam M, Payami S. High torque/ampere direct torque control of switched reluctance motor drives. In: 2019 National power electronics conference (NPEC). 2019. p. 1–6.
58. Xu A, Zhao X, He K, Cao Y. Torque-ripple reduction of SRM using optimised voltage vector in DTC. IET Electr Syst Transp. 2018;8(1):35–43. https://doi.org/10.1049/iet-est.2017.0090
59. Zhang W, Xu A, Han L, Wang S. Minimising torque ripple of SRM by applying DB-DTFC. IET Electr Power Appl. 2019;13(11):1883–90. https://doi.org/10.1049/iet-epa.2018.5441

60. Li Q, Xu A, Shang C, Zhu J, Zhou L. Non-linear deadbeat direct torque and flux control for switched reluctance motor. IET Power Electron. 2020;13(18):4176–82. https://doi.org/10.1049/iet-pel.2020.0244.
61. Deng X, Mecrow B, Wu H, Martin R. Design and development of low torque ripple variable-speed drive system with six-phase switched reluctance motors. IEEE Trans Energy Convers. 2018;33(1):420–9. https://doi.org/10.1109/TEC.2017.2753286.
62. Li C, Wang G, Li Y, Xu A. An improved finite-state predictive torque control for switched reluctance motor drive. IET Electr Power Appl. 2018;12(1):144–51. https://doi.org/10.1049/iet-epa.2017.0268.
63. Sun X, Wu J, Lei G, Guo Y, Zhu J. Torque ripple reduction of SRM drive using improved direct torque control with sliding mode controller and observer. IEEE Trans Ind Electron. 2021;68(10):9334–45. https://doi.org/10.1109/TIE.2020.3020026.
64. Rodriguez J, Garcia C, Mora A, Davari SA, Rodas J, Valencia DF, et al. Latest advances of model predictive control in electrical drives—Part II: Applications and benchmarking with classical control methods. IEEE Trans Power Electron. 2022;37(5):5047–61. https://doi.org/10.1109/TPEL.2021.3121589.
65. Valencia DF, Filho SR, Callegaro AD, Preindl M, Emadi A. Virtual-flux finite control set model predictive control of switched reluctance motor drives. In: IECON 2019—45th Annual conference of the IEEE industrial electronics society. 2019. p. 1465–70.
66. Li X, Shamsi P. Adaptive model predictive current control for DSSRM drives. In: 2014 IEEE Transportation electrification conference and expo (ITEC). 2014. p. 1–5.
67. Mikail R, Husain I, Sozer Y, Islam MS, Sebastian T. A fixed switching frequency predictive current control method for switched reluctance machines. IEEE Trans Ind Appl. 2014;50(6):3717–26. https://doi.org/10.1109/TIA.2014.2322144.
68. Mehta S, Kabir MA, Husain I. Extended speed current profiling algorithm for low torque ripple SRM using model predictive control. In: 2018 IEEE Energy conversion congress and exposition (ECCE). 2018. p. 4558–63.
69. Valencia DF, Tarvirdilu-Asl R, Garcia C, Rodriguez J, Emadi A. A review of predictive control techniques for switched reluctance machine drives. Part I: Fundamentals and current control. IEEE Trans Energy Convers. 2021;36(2):1313–22. https://doi.org/10.1109/TEC.2020.3047983
70. Peyrl H, Papafotiou G, Morari M. Model predictive torque control of a Switched Reluctance Motor. In: 2009 IEEE International conference on industrial technology. 2009. p. 1–6.
71. Villegas J, Vazquez S, Carrasco JM, Gil I. Model predictive control of a switched reluctance machine using discrete space vector modulation. In: 2010 IEEE International symposium on industrial electronics. 2010. p. 3139–44.
72. Fang G, Ye J, Xiao D, Xia Z, Emadi A. Computational-efficient model predictive torque control for switched reluctance machines with linear-model-based equivalent transformations. IEEE Trans Ind Electron. 2022;69(6):5465–77. https://doi.org/10.1109/TIE.2021.3091918.
73. Song S, Hei R, Ma R, Liu W. Model predictive control of switched reluctance starter/generator with torque sharing and compensation. IEEE Trans Transport Electrification. 2020;6(4):1519–27. https://doi.org/10.1109/TTE.2020.2975908.
74. Ren P, Zhu J, Jing Z, Guo Z, Xu A. Minimization of torque ripple in switched reluctance motor based on MPC and TSF. IEEJ Trans Electr Electron Eng. 2021;16(11):1535–43. https://doi.org/10.1002/tee.23458.
75. Shang C, Xu A, Huang L, Chen J. Flux linkage optimization for direct torque control of switched reluctance motor based on model predictive control. IEEJ Trans Electr Electron Eng. 2019;14(7):1105–13. https://doi.org/10.1002/tee.22906.
76. Li J, Ding W, Yuan J, Liu Z, Hu R. An improved model predictive control method of switched reluctance motor based on direct torque control. In: 2021 24th International conference on electrical machines and systems (ICEMS 2021). 2021. p. 2568–72.
77. Song S, Liu J, Zhao Y, Ge L, Ma R, Liu W. High-dynamic four-quadrant speed adjustment of switched reluctance machine with torque predictive control. IEEE Trans Ind Electron. 2022;69(8):7733–43. https://doi.org/10.1109/TIE.2021.3108707.

78. Sun X, Zhu Y, Cai Y, Yao M, Sun Y, Lei G. Optimized-sector-based model predictive torque control with sliding mode controller for switched reluctance motor. IEEE Trans Energy Convers. 2024;39(1):379–88. https://doi.org/10.1109/TEC.2023.3301000.
79. Valencia DF, Tarvirdilu-Asl R, Garcia C, Rodriguez J, Emadi A. A review of predictive control techniques for switched reluctance machine drives. Part II: Torque control, assessment and challenges. IEEE Trans Energy Convers. 2021;36(2):1323–35. https://doi.org/10.1109/TEC.2020.3047981
80. Kiani M. Model predictive control of stator currents in switched reluctance generators. In: 2014 IEEE 23rd International symposium on industrial electronics (ISIE). 2014. p. 842–6.
81. Husain T, Elrayyah A, Sozer Y, Husain I. Unified control for switched reluctance motors for wide speed operation. IEEE Trans Ind Electron. 2019;66:3401–11.
82. Yao S, Zhang W. A simple strategy for parameters identification of SRM direct instantaneous torque control. IEEE Trans Power Electron. 2018;33:3622–30.
83. Sun X, Feng L, Diao K, Yang Z. An improved direct instantaneous torque control based on adaptive terminal sliding mode for a segmented-rotor SRM. IEEE Trans Ind Electron. 2021;68:10569–79.
84. Sun Q, Wu J, Gan C. Optimized direct instantaneous torque control for SRMs with efficiency improvement. IEEE Trans Ind Electron. 2021;68:2072–82.
85. Li Z, Kong W, Cheng Y, Qu R. An iterative learning based direct torque control strategy of dc-biased vernier reluctance machines for torque ripple reduction. IEEE Trans Power Electron. 2023;38:15456–66.
86. Zhao X, Xu A, Zhang W. Research on dtc system with variable flux for switched reluctance motor. CES Trans Elect Mach Syst. 2017;1:199–206.
87. Yan N, Cao X, Deng Z. Direct torque control for switched reluctance motor to obtain high torque–ampere ratio. IEEE Trans Ind Electron. 2019;66:5144–52.
88. Shah V, Alam M, Payami S. High torque/ampere direct torque control of switched reluctance motor drives. 2019 National Power Electronics Conference (NPEC)2019. p. 1–6.
89. Xu A-d, Zhao X, He K, Cao Y. Torque-ripple reduction of SRM using optimised voltage vector in dtc. IET Elect Syst Transp. 2018;8:35–43.
90. Zhang W, Xu Ad, Han L, Wang S. Minimising torque ripple of SRM by applying db-dtfc. IET Electr Power Appl. 2019;13:1883–90.
91. Li Q, Xu A, Shang C, Zhu J, Zhou L. Non-linear deadbeat direct torque and flux control for switched reluctance motor. IET Power Electron. 2020;13:4176–82.

Chapter 5
Optimization Methods for Electrical Machines

Abstract Implementing multi-objective optimization of electrical machines faces challenges, such as complexity in handling conflicting objectives and constraints, computational intensity, achieving a balance between convergence and diversity, selecting an optimal solution from the Pareto front, handling high-dimensional design spaces, and making informed decisions. Optimization techniques are of paramount importance in efficiently solving these problems. This chapter provides a concise overview of optimization techniques for electrical machines. It highlights the importance of the design of experiments, sensitivity analysis, surrogate models, and multi-objective optimization algorithms in enhancing machine performance. These techniques facilitate efficient design space exploration, identification of critical parameters, approximation of complex simulations, and simultaneous optimization of multiple objectives. Implementing these optimization techniques expedites the development of advanced electrical machines with improved efficiency, reliability, and performance.

Keywords Design of experiments · Multi-objective optimization algorithms · Sensitivity analysis · Surrogate models · Electrical machines · Switched reluctance motors

5.1 Introduction

Optimization techniques play a vital role in improving the performance of electrical machines. Several interconnected techniques, including design of experiments (DoE), sensitivity analysis, surrogate models, and multi-objective optimization algorithms, are utilized to address the complexities of electrical machine design [1, 2].

The DoE serves as a foundation for optimization by systematically varying input parameters and observing their impact on performance metrics. DoE provides valuable insights into the relationships between design variables and system responses, enabling engineers to explore the design space efficiently.

X. Sun et al., *Multi-objective Design Optimization of Switched Reluctance Motor Drive Systems*, https://doi.org/10.1007/978-981-96-0672-6_5

Sensitivity analysis methods complement DoE by quantifying the influence of input parameter variations on performance metrics. Engineers can optimize these influential factors and improve machine designs by identifying critical variables through sensitivity analysis.

Surrogate models act as approximations of computationally expensive simulation or experimental models. These models are built based on a limited number of evaluations, reducing the overall computational cost. Surrogate models enable engineers to explore the design space more efficiently, rapidly evaluate objective functions, and guide the optimization process [3, 4].

Multi-objective optimization algorithms accommodate the inherent trade-offs in electrical machine design. These algorithms consider conflicting objectives, such as maximizing power output while minimizing cost or losses, and search for optimal solutions along the Pareto front. By exploring the trade-offs between objectives, multi-objective optimization algorithms provide engineers with a comprehensive understanding of the design space, allowing for informed decision-making [5–8].

The relationship between these optimization techniques is intertwined. DoE helps efficiently explore the design space, sensitivity analysis identifies critical variables within that space, surrogate models facilitate faster evaluations of objective functions, and multi-objective optimization algorithms aid in finding optimal solutions considering conflicting objectives. Together, these techniques enable engineers to navigate the complex landscape of electrical machine design and drive improvements in performance, efficiency, and reliability [9, 10].

5.2 Optimization Method Classification

Optimization methods could have several forms of classification according to different criteria. In terms of the number of objectives, they could be classified into single and multi-objective optimizations. According to the involvement of the control systems and variables, they could be classified into component and system-level optimization. Moreover, in terms of consideration of uncertainties, there are deterministic and robust optimization methods.

5.2.1 Single-Objective and Multi-objective Optimizations

The definition of the optimization problem is a primary and significant step in optimization. It includes the objectives, constraints, parameters, and boundaries. The optimization model of single-objective optimization problems can be defined as

$$\begin{aligned}
&\min : f(\mathbf{x}_s) \\
&\text{s.t.} \quad g_i(\mathbf{x}_s) \leq 0, i = 1, 2, \ldots, n \\
&\qquad \mathbf{x}_{sl} \leq \mathbf{x}_s \leq \mathbf{x}_{su}
\end{aligned} \tag{5.1}$$

where $\mathbf{x}_s$, f, and g_i are the design parameter vector, objectives, and constraints of the design candidates, respectively; $\mathbf{x}_s$ may consist of motor and control parameters, and $\mathbf{x}_{sl}$ and $\mathbf{x}_{su}$ are the lower and upper boundaries, respectively. The objective function may investigate one performance indicator of the motor, like efficiency, or a combination of several indicators with various weighting factors, such as efficiency and torque density.

However, the main problem of single-objective optimization is that it needs to run again if the weighting factors vary. It cannot provide much information to help people decide on industrial applications. If only one performance indicator is considered in the objective, it ignores and potentially influences the other important indices. Thus, multi-objective optimization problems should be taken into consideration. Generally, a multi-objective optimization model with m objectives and n constraints can be defined as

$$\begin{aligned}
&\min : \{f_1(\mathbf{x}_s), f_1(\mathbf{x}_s), \ldots, f_m(\mathbf{x}_s)\} \\
&\text{s.t.} \quad g_i(\mathbf{x}_s) \leq 0, i = 1, 2, \ldots, n. \\
&\qquad \mathbf{x}_{sl} \leq \mathbf{x}_s \leq \mathbf{x}_{su}
\end{aligned} \tag{5.2}$$

5.2.2 Component Optimization and System-Level Optimization

The operation performances of electrical machines contain two aspects, i.e., the steady-state and dynamic performances. To achieve good dynamic performance, both the electrical machines and the cooperated control systems should be evaluated. For applications like electric vehicles (EVs) or hybrid EVs, the energy efficiency of the whole system is more of a concern than just the motor. System-level design optimization puts effort into selecting the best solution for the system, which has provided a promising topic in recent years [11, 12]. The system-level optimization method for the electrical machine was first proposed in [13]. The parameters of the motor and control are optimized and evaluated during the same process. It should be noted that the optimal design solution of system-level optimization aims to find the optimal system performance, but it cannot guarantee the optimal component performance like the optimal motor topology.

5.2.3 Deterministic and Robust Optimization

The deterministic optimization approach aims to provide an optimal solution or a Pareto front for electrical machines and drive systems without considering uncertainties like material diversities, manufacturing tolerances, and assembly errors. It is acknowledged that manufactured products will feature inevitable deviations from the ideal model. Implementing the deterministic approach results in potentially dissatisfying variations of output performance, especially in mass production [14].

Figure 5.1 presents an illustration of deterministic and robust approaches. Assume that the goal of the optimization is to minimize $f(x)$. It can be observed from Fig. 5.1a that the deterministic approach follows the overall best solution. However, when the parameter x is subject to variation within the range of Δx, the optimization objective's fluctuation of the deterministic approach is significantly larger than that of the robust approach. Moreover, some variations of solutions for the deterministic approach may exceed the constraints and become infeasible, which is forbidden in industrial applications, and the rejections will increase the manufacturing cost. Figure 5.1b shows that the robust approach achieves more concentrated design solutions within the feasible domain. In contrast, the designs of the deterministic approach are usually distributed close to the constraint boundary.

The reliability and quality cannot be guaranteed using the deterministic approach. In [1], a case study for a permanent magnet motor optimization was investigated, and it has been found that the reliabilities of the motor obtained through robust optimization are 100% considering the preset constraints, while those of the deterministic designs can be lower than 30%.

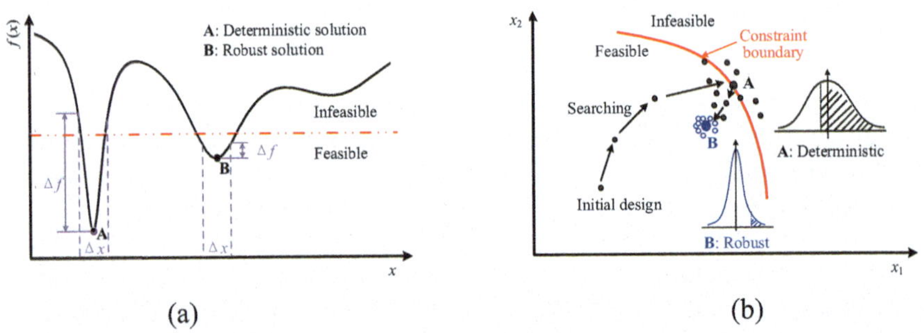

Fig. 5.1 Illustrations of deterministic **a** and robust **b** optimization approaches

5.3 Design of Experiments

DoE is a systematic approach to planning and conducting experiments, which allows researchers to determine the relationship between input variables (factors) and output responses efficiently and effectively. Fisher RA first introduced this method in the early twentieth century, and it has since become an essential tool in scientific research and industrial applications [15].

DoE involves a structured process for designing and analyzing experiments, which typically involves identifying factors that may impact the response variable, creating a design matrix that specifies the levels of each factor to be tested, conducting the experiment according to the design, and then using statistical methods to analyze the results. By systematically varying the factors of interest, researchers can identify the factors that have the most significant impact on the response variable and determine how these factors interact.

One of the key advantages of DoE is that it allows researchers to conduct experiments using a limited number of trials while still obtaining statistically valid results [16]. DoE techniques have been extensively used in the optimization of electrical machines. The most used DoE techniques include the full factorial design, fractional factorial design, response surface methodology, and the Taguchi method.

(a) Full factorial design

The full factorial design is a technique that involves testing all possible combinations of the factors that influence the performance of the electrical machine. For example, if there are three factors (such as rotor diameter, stator length, and magnet strength), each with two levels (high and low), then the full factorial design would involve testing all eight possible combinations of these factors [17, 18].

Full factorial designs are useful for studying multiple factors' main effects and interactions. By testing all possible combinations of the levels of the factors, a full factorial design allows for the identification of the impact of each factor on the response variable and the interactions among the factors. Additionally, full factorial designs can be used to optimize the response variable by identifying the optimal levels of the factors. However, a major limitation of full factorial designs is that they can be resource-intensive and time-consuming, especially when many factors and/or levels are being studied.

(b) Fractional factorial design

Fractional factorial design is an experimental design that involves testing a subset of the possible level combinations of the studied factors. This method is used when it is not feasible or practical to test all possible combinations. Fractional factorial designs are based on the concept of confounding, which involves combining two or more factors so that their effects on the response variable cannot be distinguished. In a fractional factorial design, certain combinations of factors are intentionally confounded to reduce the number of experiments required while still allowing for identifying the main effects and some of the interactions [19, 20].

The main advantage of fractional factorial designs is that they require fewer experiments than full factorial designs, which can save time and resources. However, the confounding of certain combinations of factors means that the effects of these factors cannot be estimated separately, limiting the amount of information that can be obtained from the experiment.

(c) Response surface methodology

The response surface methodology involves creating a mathematical model that describes the electrical machine's relationship between the factors and the response (performance) [18, 21]. The model is then used to predict the response variable at different levels of the factors and to identify the optimal levels of the factors for the desired response. The mathematical model is typically represented as a polynomial equation with one or more independent variables (factors) and coefficients representing the main effects and interactions among the factors.

The model predicts the optimal settings for the factors that will result in the best performance. Response surface methodology is commonly used in experimental design to improve the efficiency and effectiveness of experiments by reducing the number of experiments required to identify the optimal conditions for a desired response [18, 22].

(d) Taguchi method

Taguchi method is a statistical approach that aims to optimize the performance of the electrical machine by minimizing the effect of external factors (noise). This technique involves testing a small number of experiments with different combinations of factors and then analyzing the results to determine the optimal settings for the factors [23]. The Taguchi method differs from other DoE techniques in emphasizing the importance of designing for a target value rather than minimizing variation. This approach is based on the idea that the cost of variation can be higher than the cost of achieving a specific target value, especially regarding quality characteristics that affect customer satisfaction [12].

One of the key features of the Taguchi method is the use of orthogonal arrays, which are experimental designs that allow for identifying the optimal levels of the factors with a minimum number of experiments [12]. This approach can save time and resources compared to other DoE techniques that may require a large number of experiments.

Overall, DoE provides a powerful approach to understanding the relationship between factors and responses. It can be applied in a wide range of fields, including engineering, biology, chemistry, and social sciences. Its effectiveness and efficiency have made it an indispensable tool for researchers and practitioners.

5.4 Sensitivity Analysis Methods

Sensitivity analysis is a technique used to assess the impact of changes in certain variables on the outcomes of a system, model, or decision. It is a powerful tool for understanding the behavior of a mathematical model or simulation. It involves systematically varying the model's input parameters and observing the corresponding changes in the output. Sensitivity analysis can help researchers identify the most influential input parameters in a model or simulation and prioritize efforts to improve the accuracy and reliability of the model. It can also help identify areas of uncertainty and guide additional data collection for model calibration and validation. Sensitivity analysis is widely used in various fields, including engineering, finance, environmental science, and epidemiology, to name a few [24].

Sensitivity analysis can be classified into different categories based on various criteria. Here is a classification of sensitivity analysis based on different perspectives.

(a) Based on input variation

One-way sensitivity analysis: Examines the impact of varying one input variable at a time while keeping others constant.

Multi-way sensitivity analysis: Considers the simultaneous variation of multiple input variables to observe their combined effects and interactions.

(b) Based on analysis scope

Local sensitivity analysis: Analyzes the model's sensitivity to changes in parameters around a specific point or operating condition.

Global sensitivity analysis: Assesses the overall sensitivity of the model to variations in input parameters across their entire ranges, providing insights into the relative importance of variables and their interactions.

(c) Based on methodology

Deterministic sensitivity analysis: Focuses on analyzing the model's behavior by considering deterministic changes in input variables.

Probabilistic sensitivity analysis: Incorporates probability distributions for input variables to assess the impact of uncertainty on the model's outcomes.

(d) Based on an analytical approach

Variance-based sensitivity analysis: Quantifies the contribution of each input variable and their interactions to the overall variability of the output, using methods such as Sobol indices or analysis of variance (ANOVA).

Screening methods: Identify the most influential factors among a large set of input variables using a subset of runs or a sampling strategy.

Gradient-based methods: Utilize the model's gradients or derivatives to measure the output's sensitivity with respect to input variables.

(e) Based on simulation technique

Monte Carlo sensitivity analysis: Involves generating random samples from input distributions and running the model or simulation for each sample to assess the sensitivity to input variations.

Latin hypercube sampling: A technique that optimizes sampling efficiency by stratifying the input variable ranges.

(f) Based on objective

Parameter screening: Identifies the most influential parameters that significantly affect the model's outputs to reduce the dimensionality of the analysis.

Uncertainty analysis: Evaluates the impact of uncertainty in input variables on the model's outputs, often in conjunction with sensitivity analysis.

These classifications provide different perspectives on sensitivity analysis, and the choice of approach depends on the specific context, objectives, available data, and computational resources. Among them, local sensitivity analysis and global sensitivity analysis will be introduced in the following sections.

5.4.1 Local Sensitivity Analysis

Local sensitivity analysis is a powerful technique widely employed in research and engineering to assess the sensitivity of mathematical models or systems to small perturbations or variations in input parameters around a specific operating point. It provides valuable insights into the local behavior and stability of the system, offering a quantitative understanding of how changes in input variables near the chosen point affect the model's outputs or outcomes [25, 26].

At the core of local sensitivity analysis lie the sensitivity coefficients, which represent the partial derivatives of the output variables with respect to the input parameters. The sensitivity coefficient S_{ij} quantifies the effect of a small change in the jth input parameter on the ith output variable at the specified operating point. These coefficients serve as quantitative measures of the local sensitivity, shedding light on the relative importance and influence of different input variables on the system's behavior.

One approach to conducting local sensitivity analysis is to derive the model equations and calculate the sensitivities directly analytically. The sensitivity coefficients can be obtained by computing the partial derivatives of each output variable with respect to each input parameter. This analytical approach is suitable when the model equations are known and amenable to mathematical manipulation. For instance, let us consider a model with n input parameters, $X_1, X_2, \ldots, X_n$, and m output variables, $Y_1, Y_2, \ldots, Y_m$. The sensitivity coefficient S_{ij} can be expressed as

$$S_{ij} = \frac{\partial Y_i}{\partial X_j}. \tag{5.3}$$

However, in cases where the model equations are complex or not readily available, numerical methods such as finite difference approximations can be employed to estimate the sensitivity coefficients. These methods involve introducing small perturbations in the input parameters to estimate the resulting changes in the output variables. By comparing the output values, the sensitivity coefficients can be estimated. For example, using the finite difference method, the sensitivity coefficient S_{ij} can be approximated as

$$S_{ij} = \frac{f_i(X_1, X_2, \ldots, X_j + \Delta X_j, \ldots, X_n) - f_i(X_1, X_2, \ldots, X_j - \Delta X_j, \ldots, X_n)}{2\Delta X_j}. \tag{5.4}$$

Here, ΔX_j represents a small perturbation in the jth input parameter.

The results obtained from local sensitivity analysis provide critical insights into the system's sensitivity to changes in input variables around the chosen operating point. The magnitudes and signs of the sensitivity coefficients indicate the strength and direction of the influence, respectively. By identifying the most influential input parameters and their relative importance in affecting the output variables, researchers gain a deeper understanding of the local behavior and stability of the system. This knowledge empowers researchers to make informed decisions, optimize processes, and refine the model to enhance its predictive capabilities.

Local sensitivity analysis finds applications in various academic disciplines, including environmental science, economics, engineering, biology, healthcare, etc. It is crucial in sensitivity assessment, uncertainty quantification, risk analysis, and decision support systems. Moreover, by providing localized insights into the model's behavior, local sensitivity analysis enhances our understanding of complex systems and aids in developing robust and reliable models [27].

5.4.2 *Global Sensitivity Analysis*

Global sensitivity analysis is a powerful methodology employed in academic research and engineering to assess the sensitivity of mathematical models or systems to variations in input parameters across their entire range. Unlike local sensitivity analysis, which focuses on a specific operating point, global sensitivity analysis provides a comprehensive understanding of how changes in input variables influence the model's outputs or outcomes across the entire parameter space. This information is crucial for identifying the most influential parameters, understanding the system's behavior, and making robust decisions [28].

The foundation of global sensitivity analysis lies in quantifying the contributions of individual input parameters or combinations of parameters to the variability in the model outputs. Various techniques have been developed to measure global sensitivity, and one widely used approach is based on variance decomposition. This approach assesses the total variance of the model output and decomposes it into components attributed to individual parameters or their interactions.

One commonly used global sensitivity analysis method is Sobol indices. The first-order Sobol index (S_i) measures the fraction of the total output variance attributed to the variation in the ith input parameter alone. It is defined as [11, 28]

$$S_i = \frac{\text{var}_{X_i}\left(E_{X_{\sim i}}(Y)\right)}{\text{var}(Y)} \tag{5.5}$$

where $\text{var}_{Xi}(E_{X\sim i}(Y))$ represents the variance of the conditional expectation of the output Y given all input variables except X_i, and var(Y) is the total variance of the output. The first-order Sobol index quantifies the individual contribution of each parameter to the output variance, indicating its relative importance.

Additionally, the total-order Sobol index (S_{Ti}) measures the fraction of the total output variance attributed to the variation in the ith input parameter, including its interactions with other parameters. It can be calculated as

$$S_{Ti} = \frac{\text{var}_{X_i}(E_{X\sim i}(Y)) + \sum_{j\neq i} \text{var}_{X_i X_j}\left(E_{X_{\sim i,j}}(Y)\right)}{\text{var}(Y)} \tag{5.6}$$

where $\text{var}_{XiXj}(\text{E}_{X\sim i,j}(Y))$ represents the variance of the conditional expectation of the output Y given all input variables except X_i and X_j, and the summation term accounts for the contributions of interactions between the ith parameter and all other parameters. The total-order Sobol index captures the total impact of a parameter, including its direct effect and interactions, on the output variance.

The results of global sensitivity analysis help researchers understand the relative importance of input parameters, identify influential parameters, and quantify their interactions. Global sensitivity analysis facilitates robust decision-making, model optimization, and uncertainty quantification by focusing on the entire parameter space. It allows researchers to prioritize resources and efforts toward the most critical parameters, leading to more effective model calibration, validation, and improvement.

5.5 Surrogate Models

Surrogate models, also known as metamodels, are simplified mathematical or statistical models that approximate the behavior of complex systems or processes. These surrogate models are constructed using a smaller set of data or simulations from the original system and are designed to mimic the response of the original system with reasonable accuracy [4, 29].

The primary purpose of surrogate models is to provide a computationally efficient alternative to the computationally expensive or time-consuming original models.

They can be used in various fields, including engineering, optimization, computer simulations, and data analysis, where the original models might be too slow or impractical to use directly.

The construction of a surrogate model typically involves selecting a suitable mathematical or statistical form, such as polynomial regression, neural networks, Gaussian processes, or support vector machines, and then training the model using available data points or simulation results from the original system. The surrogate model is calibrated to minimize the discrepancy between its predictions and the outputs of the original model at the selected data points.

Once the surrogate model is constructed, it can approximate the original system's behaviour for new input values or scenarios. Using the surrogate model can reduce the computational burden of running the original model repeatedly or exploring a large parameter space [30]. Several types of surrogate models can be used in the design optimization of electrical machines.

5.5.1 Polynomial Regression Models

Polynomial regression is a type of regression analysis in which the relationship between the independent variable (input) and the dependent variable (output) is modeled as an nth-degree polynomial. It is an extension of simple linear regression, allowing for more complex relationships between the variables [31].

The polynomial regression model can be represented as

$$y = \beta_0 + \beta_1 x + \beta_2 x^2 + \cdots + \beta_n x^n + \varepsilon \tag{5.7}$$

where x is the independent (input) variable, y is the dependent (output) variable, $\beta_0, \beta_1, \ldots, \beta_n$ are the coefficients (parameters) of the polynomial regression model that need to be estimated, $x^2, x^3, \ldots, x^n$ are the higher-order terms of x, raised to the powers 2, 3, …, n, respectively, and ε represents the error term or the residuals, which account for the difference between the predicted values and the actual values. Polynomial regression aims to estimate the coefficients $\beta_0, \beta_1, \ldots, \beta_n$ that best fit the data. This is typically done by minimizing the sum of squared residuals, which measures the overall discrepancy between predicted and actual values.

In matrix form, the polynomial regression equation can be expressed as

$$\mathbf{Y} = \mathbf{X}\boldsymbol{\beta} + \varepsilon \tag{5.8}$$

where $\mathbf{X}$ is the design matrix that includes the polynomial terms of the independent variable(s), $\mathbf{Y}$ is a vector of observed values for the dependent variable, and $\boldsymbol{\beta}$ is a vector of coefficients to be estimated. The coefficients $\boldsymbol{\beta}$ in the polynomial regression equation can be estimated using the method of ordinary least squares (OLS). The OLS estimator for $\boldsymbol{\beta}$ is given by

$$\boldsymbol{\beta} = \left(\mathbf{X}^T\mathbf{X}\right)^{-1}\mathbf{X}^T\mathbf{Y}. \tag{5.9}$$

Various approaches can be employed to determine the optimal polynomial degree. One common method is cross-validation, where the dataset is divided into training and validation sets. The polynomial degree that minimizes the validation error is selected as the optimal degree.

Residuals represent the differences between the observed and predicted values obtained from the polynomial regression model. The residual for each observation can be calculated as

$$\varepsilon = \mathbf{Y} - \hat{\mathbf{Y}} \tag{5.10}$$

where $\hat{\mathbf{Y}}$ is the vector of predicted values obtained from the polynomial regression model.

Several measures can assess the goodness-of-fit of the polynomial regression model. The coefficient of determination (R-squared) is commonly used and is calculated as the proportion of the total sum of squares (SST) explained by the regression model, which can be calculated by

$$R^2 = 1 - (\mathrm{SSR}/\mathrm{SST}) \tag{5.11}$$

where SSR is the sum of squared residuals, and SST is the total sum of squares.

The SSR equation measures the variability in the dependent variable that the regression model explains. It can be calculated as

$$\mathrm{SSR} = \sum \left(\varepsilon^2\right) = \sum \left(\mathbf{Y} - \hat{\mathbf{Y}}\right)^2. \tag{5.12}$$

The SST equation quantifies the total variability in the dependent variable around its mean. It is given by

$$\mathrm{SST} = \sum \left(\mathbf{Y} - \overline{\mathbf{Y}}\right)^2 \tag{5.13}$$

where $\overline{\mathbf{Y}}$ is the mean of the dependent variable.

5.5.2 Radial Basis Function Network Models

Radial basis function (RBF) network models are machine learning models that use radial basis functions as activation functions. These models are often used for regression, classification, and clustering tasks. The basic idea behind RBF models is to transform input data into a high-dimensional feature space using radial basis functions. A radial basis function is a mathematical function that depends only on the distance from the origin or a center point. Commonly used radial basis functions include the Gaussian and inverse multiquadric functions [32].

The RBF model typically consists of three main components, as the following.

Input layer: The input layer receives the input data and passes it to the hidden layer.

Hidden layer: The hidden layer applies radial basis functions to transform the input data into a higher-dimensional feature space. Each node in the hidden layer represents a radial basis function centered at a specific point in the input space. The activation of each node is computed based on the distance between the input data and the center of the radial basis function.

Output layer: The output layer performs the final computation based on the activations of the hidden layer nodes. For regression tasks, the output layer may use a linear activation function. The output layer may employ a softmax function for classification tasks to produce probabilities for different classes.

The radial basis function $\varphi_i(x)$ can be expressed as

$$\varphi_i(x) = \exp\left(-\gamma_i * \|x - c_i\|^2\right) \tag{5.14}$$

where $\varphi_i(x)$ is the activation of the ith radial basis function, $\|x - c_i\|$ is the distance between input vector x and center c_i, and γ_i is the width parameter. The hidden layer consists of k nodes, each representing a different radial basis function. The activations of these basis functions form a vector:

$$\phi(x) = [\varphi_1(x), \varphi_2(x), \ldots, \varphi_k(x)]. \tag{5.15}$$

The above expression represents a commonly used radial basis function known as the Gaussian radial basis function or the Gaussian kernel. In an RBF model, multiple radial basis functions are typically used, each centered at a different point in the input space. The activations of these basis functions are computed simultaneously, and their weighted sum forms the output of the hidden layer.

5.5.3 *Kriging Models*

Kriging models, also known as Gaussian process regression, are a type of statistical interpolation technique used to estimate values at unobserved locations based on observations at known locations. They are commonly employed in geostatistics and spatial analysis.

Kriging models assume that the estimated variable has spatial correlation or dependence, meaning nearby locations are more similar than farther apart locations. The models utilize the concept of semivariance or covariance to quantify the degree of spatial dependence. The semivariogram or covariance function is typically calculated from the observed data, and it describes how the similarity between values decreases with increasing spatial separation [33].

There are several types of kriging models, each with its own assumptions and characteristics. Here are some commonly used variants:

Ordinary Kriging: Assumes a constant mean value and stationary spatial dependence. It provides the best linear unbiased estimate of the variable at unobserved locations.
Universal Kriging: Similar to Ordinary Kriging, it incorporates a trend component into the model. It is useful when the variable's mean value varies across the study area.
Simple Kriging: Assumes a known constant mean value. It is suitable when the mean is known or can be estimated accurately.

The equations for Kriging models involve calculating the weights assigned to the observed data points and combining them to estimate the value at an unobserved location. The Ordinary Kriging, Universal Kriging, and Simple Kriging equations are presented below.

(a) Ordinary Kriging [34]:

The estimated value at the unobserved location z(x) can be calculated as

$$z(x) = \hat{\mu} + \sum ([\lambda(i) * (z(i) - \hat{\mu})]) \tag{5.16}$$

where $z(x)$ is the estimated value at location x, $\hat{\mu}$ the estimated mean of the variable, and $\lambda(i)$ the weight assigned to each observed value $z(i)$ at location i. The weights $\lambda(i)$ are determined by minimizing the kriging variance, subject to the unbiasedness constraint:

$$\sum \lambda(i) = 1. \tag{5.17}$$

The Kriging variance is given by

$$\mathrm{var}[z(x)] = \sum [\lambda(i) * \mathrm{cov}(z(i), z(x)) + \lambda(0)] \tag{5.18}$$

where $\mathrm{cov}(z(i), z(x))$ is the covariance between the observed value at location i and the estimated value at location x, $\lambda(0)$ is the Lagrange multiplier, which ensures the unbiasedness constraint is satisfied.

(b) Universal Kriging [35]:

In Universal Kriging, a trend component is incorporated into the model. The equation for estimating the value at an unobserved location $z(x)$ is

$$z(x) = \mu(x) + \sum ([\lambda(i) * (z(i) - \mu(i))]) \tag{5.19}$$

where $\mu(x)$ is the estimated trend component at location x, $\mu(i)$ is the estimated trend component at the observed location i, and the other terms have the same meanings as in Ordinary Kriging.

(c) Simple Kriging [36]:

In Simple Kriging, a known constant mean value is assumed. The equation for estimating the value at an unobserved location $z(x)$ is

$$z(x) = \mu + \sum([\lambda(i) * (z(i) - \mu)]) \tag{5.20}$$

where μ is the known mean value, and the other terms have the same meanings as in Ordinary Kriging.

These equations represent the general form of kriging models. The specific calculations of the weights and covariances depend on the choice of semivariogram or covariance model, which describes the spatial dependence structure of the variable being estimated.

5.5.4 *Support Vector Regression*

Support Vector Regression (SVR) is a machine-learning algorithm for regression tasks. It is an extension of Support Vector Machines (SVMs), primarily used for classification. SVR aims to build a regression model by finding the best possible hyperplane that maximizes the margin while minimizing the error between the predicted and actual values.

SVR aims to find a hyperplane that best fits the training data while allowing error tolerance. The SVR model can be expressed as follows [37]:

Given a training set of N samples, the SVR model seeks to find the function $f(x)$ that approximates the actual output values y. The SVR model can be defined as

$$f(x) = w \cdot \phi(x) + b \tag{5.21}$$

where $f(x)$ is the predicted output value for input x, w is the weight vector, $\phi(x)$ is the feature mapping function that transforms the input data x into a higher-dimensional feature space, and b is the bias term.

SVR uses a loss function to measure the error between predicted and actual output. The loss function aims to minimize the error while maintaining a tolerance ε. The loss function used in SVR is typically the epsilon-insensitive loss function:

$$L(y, f(x)) = \max(0, |y - f(x)| - \varepsilon) \tag{5.22}$$

where y is the true output value, $f(x)$ is the predicted output value, and ε is the epsilon value that controls error tolerance.

SVR involves solving an optimization problem to find the optimal hyperplane and associated model parameters. The solution involves finding the appropriate values for the hyperparameters, such as the choice of kernel function, regularization parameter, and tolerance level. Proper selection and tuning of these hyperparameters are critical to achieving accurate and robust predictions.

5.6 Multi-objective Optimization Algorithms

Multi-objective optimization algorithms are used to find the best compromise between two or more conflicting objectives. In these problems, no single optimal solution satisfies all objectives; instead, a set of solutions, known as the Pareto front, represents the best trade-off between the objectives. The common multi-objective optimization algorithms include the non-dominated sorting genetic algorithm (NSGA) and its improved version NSGA-II.

5.6.1 Pareto Optimal Solutions

In multi-objective optimization problems, each optimization objective may be contradictory. Improving one optimization objective alone may cause performance degradation of other optimization objectives. For example, in the multi-objective optimization problem of SRMs, the output torque and loss conflict, and the increase of output torque are often accompanied by the increase of loss. Therefore, when the optimization objective is set reasonably, multi-objective optimization often cannot obtain the optimal value of each optimization objective. The final solution set is usually a compromise between various optimization objectives, and its corresponding optimal solution set is called the Pareto optimal solution set. It can be described as follows: if any optimization objective cannot be further improved without degrading the performance of other optimization objectives, then the solution is called Pareto optimal [38].

To further describe the Pareto optimal solution set, the following definition can be introduced [39]:

Definition 1: Given two vectors $\mathbf{x}, \mathbf{y} \in \mathbf{R}^k$, we say that $\mathbf{x} \leq \mathbf{y}$ if $x_i \leq y_i$ for $i = 1, \ldots, k$, and that $\mathbf{x}$ dominates $\mathbf{y}$ if $\mathbf{x} \leq \mathbf{y}$ and $\mathbf{x} \neq \mathbf{y}$.

Definition 2: We say that $\mathbf{x}$ is nondominated with respect to X if $\mathbf{x} \in X$ and there is no $\mathbf{x}' \in X$ such that $f(\mathbf{x}') < f(\mathbf{x})$.

Definition 3: If $\mathbf{x}^* \in \mathbf{F}$ and $\mathbf{x}^*$ is nondominated in $\mathbf{F}$, then $\mathbf{x}^*$ is called Pareto optimal solution.

Definition 4: The Pareto optimal set $\mathbf{P}^*$ is defined as

$$\mathbf{P}^* = \{\mathbf{x} \in \mathbf{F} | \mathbf{x} \text{ is Pareto-optimal}\}. \tag{5.23}$$

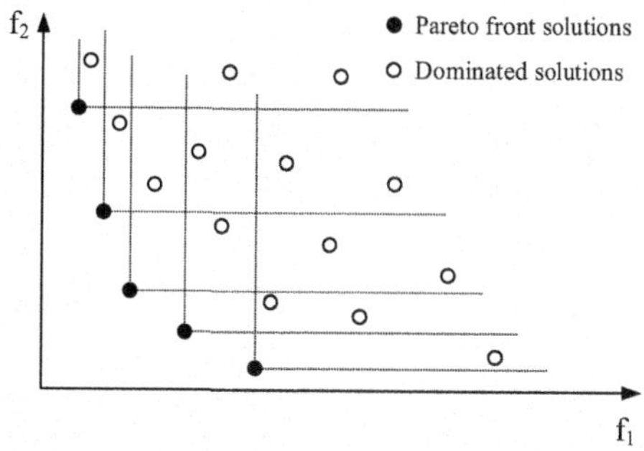

Fig. 5.2 Schematic diagram of a non-dominated solutions

Definition 5: The Pareto front **PF*** is defined as

$$\mathbf{PF}^* = \{f(\mathbf{x}) \in \mathbf{R}^k | \mathbf{x} \in \mathbf{P}^*\}. \tag{5.24}$$

Figure 5.2 depicts a schematic diagram of non-dominated solutions (Pareto optimal solutions), where all the non-dominated solutions form the Pareto front for the two objective functions shown in the figure. Combining Definition 2, for each dominated solution in the graph, another solution that outperforms the dominated solution in all optimization objectives does not exist. Conversely, for each dominated solution, another solution that outperforms the dominated solution in all optimization objectives always exists.

5.6.2 NSGA and NSGA-II

NSGA and NSGA-II are multi-objective optimization algorithms based on the principles of genetic algorithms. These algorithms are designed to solve optimization problems that involve multiple conflicting objectives. The main idea behind NSGA and NSGA-II is to maintain a population of candidate solutions called individuals, each representing a possible solution to the problem at hand. These individuals are evaluated based on their performance with respect to multiple objectives.

NSGA was initially proposed by Srinivas and Deb in 1994 as a multi-objective optimization algorithm based on genetic algorithms. It introduced the concept of nondominated sorting and elitism, allowing for the identification of Pareto-optimal solutions [40]. NSGA-II was proposed as an improvement over the original NSGA. It addressed some limitations of NSGA and introduced new features such as the crowding distance measure and a more efficient sorting mechanism. The crowding distance measure promotes diversity in the population by encouraging the maintenance of solutions with greater objective space coverage [41].

Figure 5.3 presents a flowchart of NSGA-II. This brief flowchart summarizes the main steps involved in NSGA-II. It provides a general overview of the algorithm's flow, including initialization, reproduction, environmental selection, selection for the next generation, and termination. However, for a more comprehensive understanding of the algorithm, it is recommended to refer to the original research paper by Deb et al. [41].

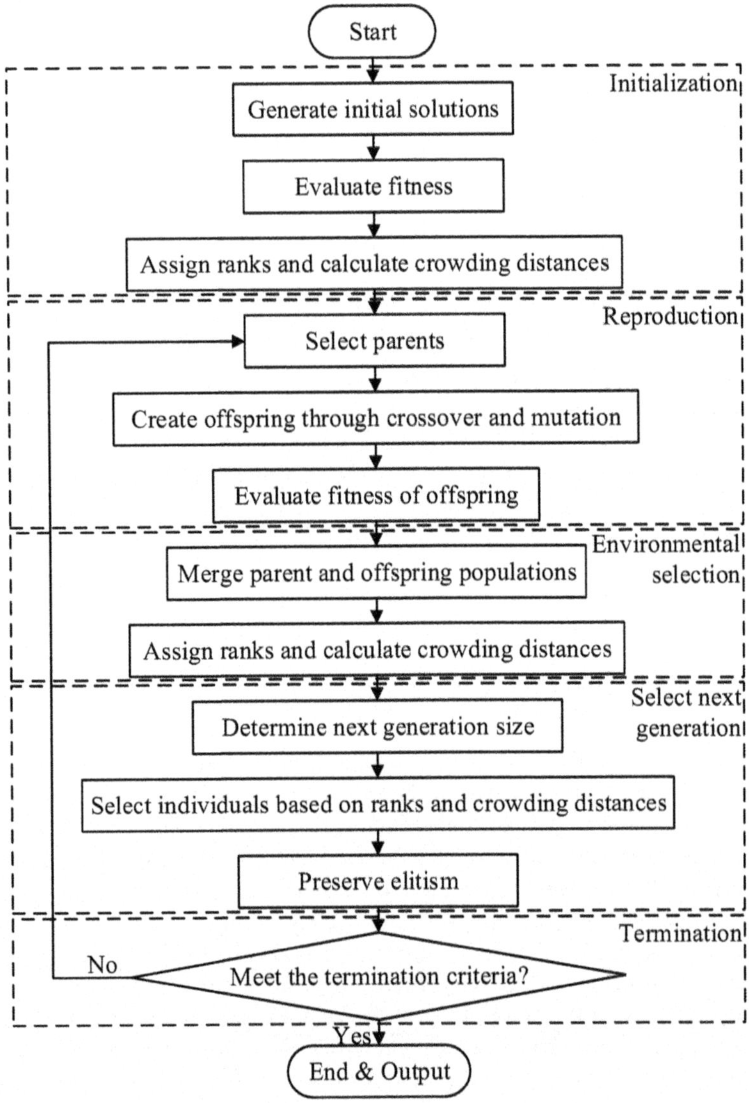

Fig. 5.3 Flowchart of NSGA-II

5.7 Summary

This chapter briefly summarizes optimization techniques used for designing and optimizing electrical machines, including DoE, sensitivity analysis, surrogate models, and multi-objective optimization algorithms. Five kinds of surrogate models, namely polynomial regression models, radial basis function network models, Kriging models, and support vector models, have been presented with details. The concept of Pareto optimal solutions for multi-objective optimization problems is illustrated, and the optimization algorithms of NSGA and NSGA-II are presented.

References

1. Lei G, Bramerdorfer G, Ma B, Guo Y, Zhu J. Robust design optimization of electrical machines: multi-objective approach. IEEE Trans Energy Convers. 2021;36(1):390–401. https://doi.org/10.1109/TEC.2020.3003050.
2. Lei G, Wang T, Zhu J, Guo Y, Wang S. System-level design optimization method for electrical drive systems—robust approach. IEEE Trans Ind Electron. 2015;62(8):4702–13. https://doi.org/10.1109/TIE.2015.2404305.
3. Easum JA, Nagar J, Werner PL, Werner DH. Efficient multiobjective antenna optimization with tolerance analysis through the use of surrogate models. IEEE Trans Antennas Propag. 2018;66(12):6706–15. https://doi.org/10.1109/TAP.2018.2870338.
4. Diao K, Sun X, Bramerdorfer G, Cai Y, Lei G, Chen L. Design optimization of switched reluctance machines for performance and reliability enhancements: a review. Renew Sust Energ Rev. 2022;168. https://doi.org/10.1016/j.rser.2022.112785
5. Öksüztepe E. In-wheel switched reluctance motor design for electric vehicles by using a pareto-based multiobjective differential evolution algorithm. IEEE Trans Veh Technol. 2017;66(6):4706–15. https://doi.org/10.1109/TVT.2016.2618119.
6. Ma C, Qu L. Multiobjective optimization of switched reluctance motors based on design of experiments and particle swarm optimization. IEEE Trans Energy Convers. 2015;30(3):1144–53. https://doi.org/10.1109/tec.2015.2411677.
7. Diao K, Bramerdorfer G, Sun X, Yang Z, Han S. Multiobjective design optimization of a novel switched reluctance motor with unequal alternating stator yoke segments. IEEE Trans Transp Electrification. 2023;9(1):512–21. https://doi.org/10.1109/tte.2022.3174337.
8. Lei G, Bramerdorfer G, Liu C, Guo Y, Zhu J. Robust design optimization of electrical machines: a comparative study and space reduction strategy. IEEE Trans Energy Convers. 2021;36(1):300–13. https://doi.org/10.1109/TEC.2020.2999482.
9. Bramerdorfer G, Tapia JA, Pyrhönen JJ, Cavagnino A. Modern electrical machine design optimization: techniques, trends, and best practices. IEEE Trans Ind Electron. 2018;65(10):7672–84. https://doi.org/10.1109/TIE.2018.2801805.
10. Lei G, Liu C, Zhu J, Guo Y. Techniques for multilevel design optimization of permanent magnet motors. IEEE Trans Energy Convers. 2015;30(4):1574–84. https://doi.org/10.1109/TEC.2015.2444434.
11. Diao K, Sun X, Lei G, Bramerdorfer G, Guo Y, Zhu J. System-level robust design optimization of a switched reluctance motor drive system considering multiple driving cycles. IEEE Trans Energy Convers. 2021;36(1):348–57. https://doi.org/10.1109/tec.2020.3009408.
12. Diao K, Sun X, Lei G, Bramerdorfer G, Guo Y, Zhu J. Robust design optimization of switched reluctance motor drive systems based on system-level sequential Taguchi method. IEEE Trans Energy Convers. 2021;36(4):3199–207. https://doi.org/10.1109/tec.2021.3085668.

13. Lei G, Wang T, Guo Y, Zhu J, Wang S. System-level design optimization methods for electrical drive systems: deterministic approach. IEEE Trans Ind Electron. 2014;61(12):6591–602. https://doi.org/10.1109/TIE.2014.2321338.
14. Bramerdorfer G. Tolerance analysis for electric machine design optimization: classification, modeling and evaluation, and example. IEEE Trans Magn. 2019;55(8):1–9. https://doi.org/10.1109/TMAG.2019.2903029.
15. Fisher RA. Design of experiments. BMJ. 1936;1(3923):554.
16. Box GE, Hunter WH, Hunter S. Statistics for experimenters. New York: Wiley; 1978.
17. Buragohain M, Mahanta C. A novel approach for ANFIS modelling based on full factorial design. Appl Soft Comput. 2008;8(1):609–25.
18. Sung-Il K, Jung-Pyo H, Young-Kyoun K, Hyuk N, Han-Ik C. Optimal design of slotless-type PMLSM considering multiple responses by response surface methodology. IEEE Trans Magn. 2006;42(4):1219–22. https://doi.org/10.1109/TMAG.2006.871950.
19. Viola J, Angel L, Sebastian JM. Design and robust performance evaluation of a fractional order PID controller applied to a DC motor. IEEE/CAA J Autom Sin. 2017;4(2):304–14. https://doi.org/10.1109/JAS.2017.7510535.
20. Ruowen R, Lowther DA, Malik Z, Hua S, Nelder J, Spence R. Applying response surface methodology in the design and optimization of electromagnetic devices. IEEE Trans Magn. 1997;33(2):1916–9. https://doi.org/10.1109/20.582665.
21. Lee JH. Optimum shape design solution of flux switching motor using response surface methodology and new type winding. IEEE Trans Magn. 2012;48(4):1637–40. https://doi.org/10.1109/tmag.2011.2173564.
22. Ishikawa T, Yamada M, Kurita N. Design of magnet arrangement in interior permanent magnet synchronous motor by response surface methodology in consideration of torque and vibration. IEEE Trans Magn. 2011;47(5):1290–3. https://doi.org/10.1109/TMAG.2010.2091399.
23. Karna SK, Sahai R. An overview on Taguchi method. Int J Eng Math Sci. 2012;1(1):1–7.
24. Iooss B, Saltelli A. Introduction to sensitivity analysis. In: Handbook of uncertainty quantification. Cham: Springer International Publishing; 2017. p. 1103–22.
25. Morio J. Global and local sensitivity analysis methods for a physical system. Eur J Phys. 2011;32(6):1577.
26. Bramerdorfer G, Zavoianu A-C. Surrogate-based multi-objective optimization of electrical machine designs facilitating tolerance analysis. IEEE Trans Magn. 2017;53(8):1–11. https://doi.org/10.1109/tmag.2017.2694802.
27. Diao K, Sun X, Lei G, Guo Y, Zhu J. Multiobjective system level optimization method for switched reluctance motor drive systems using finite-element model. IEEE Trans Ind Electron. 2020;67(12):10055–64. https://doi.org/10.1109/tie.2019.2962483.
28. Iooss B, Lemaître P. A review on global sensitivity analysis methods. In: Uncertainty management in simulation-optimization of complex systems: algorithms and applications. 2015. p. 101–22.
29. Diao K, Sun X, Lei G, Guo Y, Zhu J. Multimode optimization of switched reluctance machines in hybrid electric vehicles. IEEE Trans Energy Convers. 2021;36(3):2217–26. https://doi.org/10.1109/tec.2020.3046721.
30. Shi Z, Sun X, Lei G, Tian X, Guo Y, Zhu J. Multiobjective optimization of a five-phase bearingless permanent magnet motor considering winding area. IEEE/ASME Trans Mechatron. 2022;27(5):2657–66. https://doi.org/10.1109/TMECH.2021.3121802.
31. Ostertagová E. Modelling using polynomial regression. Procedia Eng. 2012;48:500–6.
32. Buhmann MD. Radial basis functions. Acta Numer. 2000;9:1–38.
33. Simpson TW, Mauery TM, Korte JJ, Mistree F. Kriging models for global approximation in simulation-based multidisciplinary design optimization. AIAA J. 2001;39(12):2233–41.
34. Cressie N. Spatial prediction and ordinary kriging. Math Geol. 1988;20:405–21.
35. Zimmerman D, Pavlik C, Ruggles A, Armstrong MP. An experimental comparison of ordinary and universal kriging and inverse distance weighting. Math Geol. 1999;31:375–90.
36. Daya AA, Bejari H. A comparative study between simple kriging and ordinary kriging for estimating and modeling the Cu concentration in Chehlkureh deposit, SE Iran. Arab J Geosci. 2015;8:6003–20.

37. Smola AJ, Schölkopf B. A tutorial on support vector regression. Stat Comput. 2004;14:199–222.
38. Lei G, Guo Y, Zhu J. Multidisciplinary design optimization methods for electrical machines and drive systems. Berlin: Springer; 2016.
39. Reyes-Sierra M, Coello CC. Multi-objective particle swarm optimizers: a survey of the state-of-the-art. Int J Comput Intell Res. 2006;2(3):287–308.
40. Srinivas N, Deb K. Muiltiobjective optimization using nondominated sorting in genetic algorithms. Evol Comput. 1994;2(3):221–48.
41. Deb K, Pratap A, Agarwal S, Meyarivan T. A fast and elitist multiobjective genetic algorithm: NSGA-II. IEEE Trans Evol Comput. 2002;6(2):182–97.

Chapter 6
Multi-objective Deterministic Optimization Method for Switched Reluctance Motor Drive Systems

Abstract This chapter introduces the system-level optimization method for switched reluctance motor drive systems, which optimizes the performance of an entire system rather than just individual components or subsystems. The multi-level optimization method is proposed to reduce the computational cost and solve the main obstacle of the optimization problem. A design example of an SRM drive system is presented to show system-level and multi-level optimization implementation.

Keywords Design method · Drive systems · Multi-objective optimization · Switched reluctance motors

6.1 Introduction

Motor topology development and optimization are two main aspects that improve the overall performance of SRMs. In recent years, a variety of novel structures of SRMs have been presented, such as linear SRM [1], segmented stator SRM [2], double-stator SRM [3], and axial-flux SRM [4]. Previous SRM system design and optimization research mainly focused on one objective, such as torque ripple, torque density, or efficiency. However, this single-objective optimization potentially influences the other important performance indices. Therefore, the trade-offs to the overall consideration and the implementation of multi-objective optimization are necessary to meet the requirements for different applications [5, 6].

Furthermore, previous works are mostly on the component rather than the system level [7, 8]. Theoretically, assembling individually optimized components into a system cannot guarantee optimal system performance as all parts are coupled. Thus, the perfect cooperation of motor and control levels should be optimized simultaneously [9].

The first step in the design optimization of an SRM is to build the analysis model for the performance parameters. Two criteria to evaluate the created model are accuracy and computational speed. Commonly, there are three kinds of analysis models, namely, finite element analysis (FEA) [10, 11], curve-fitting method [12],

X. Sun et al., *Multi-objective Design Optimization of Switched Reluctance Motor Drive Systems*, https://doi.org/10.1007/978-981-96-0672-6_6

and magnetic equivalent circuit (MEC) method [13]. FEA is the most popular numerical analysis method. The merit of FEA is that it can provide a comprehensive and accurate solution for nonlinear problems. However, it may not be appropriate for optimizing motors with complex structures and high dimensions due to the expensive computation cost of FEA. As an alternative, several surrogate models, such as the Kriging model [14], response surface model (RSM) [15], radial basis functions model (RBF) [16], and artificial neural network model, have been employed to reduce the FEA computation cost and improve the efficiency. As the flux linkage of the SRM is nonlinear with respect to both the current value and rotor position, the curve-fitting method approximates the flux linkage profile with mathematical expressions or intelligent models. The main drawbacks of the curve-fitting method are that they are empirical and still require data from FEA or experiments. The MEC method is analogous to an electric circuit where the reluctances of each component of the SRM are analyzed and determined. It has a fast computational speed but lacks accuracy due to its assumptions on the magnetic flux paths [17]. The second step is to find the optimal solutions for the optimization problems. In the past two decades, various intelligent algorithms, such as genetic algorithm [18], particle swarm optimization algorithm [19], and differential evolution algorithm (DEA) [20], have been implemented in the optimization process.

On the other hand, a variety of control methods have been proposed for SRMs, such as conventional current chopping control [21], angular position control (APC) [22], direct torque control [23], and model predictive control [24]. However, they are only optimized on the control level for the designed SRMs due to the largely increased number of optimization parameters at the very first stage of the design and optimization of the SRM drive system [25].

Coupling the control aspect with the electromagnetic part is beneficial to achieve the best performance of the SRM drive system. This chapter presents a novel multi-objective system-level optimization method for SRM drive systems. Both the motor and control aspects are optimized by the multi-objective optimization method. Besides, the sequential subspace optimization method is performed to reduce the computation cost and improve the optimization efficiency. The remainder of this chapter is organized as follows. Section 6.2 presents the proposed system-level optimization method. Section 6.3 presents the multi-level optimization technique. Section 6.4 investigates a case study for the design optimization of a segmented-rotor switched reluctance motor (SRSRM) and its control approach, followed by a summary in Sect. 6.5.

6.2 System-Level Optimization Method

Electrical machines are vital in various industrial applications, such as electric vehicles, renewable energy systems, and robotics. However, designing efficient electrical machines with optimized performance is a challenging task, and it requires a comprehensive understanding of the machine's behavior and operating conditions.

System-level optimization methods effectively address this challenge by integrating different design aspects into a single optimization framework [9, 26, 27].

Figure 6.1 presents a brief framework for the system-level design optimization of electrical drive systems. This kind of design optimization method considers the motor and its control system simultaneously. It can be divided into the following five steps [9].

Step 1: Determine the application requirements for the motor drive system.

The applications determine the specific requirements of the motor system, including many factors, such as rated power, rated speed, volume, and efficiency.

Step 2: According to the application requirements, select the motor and control types for the drive system.

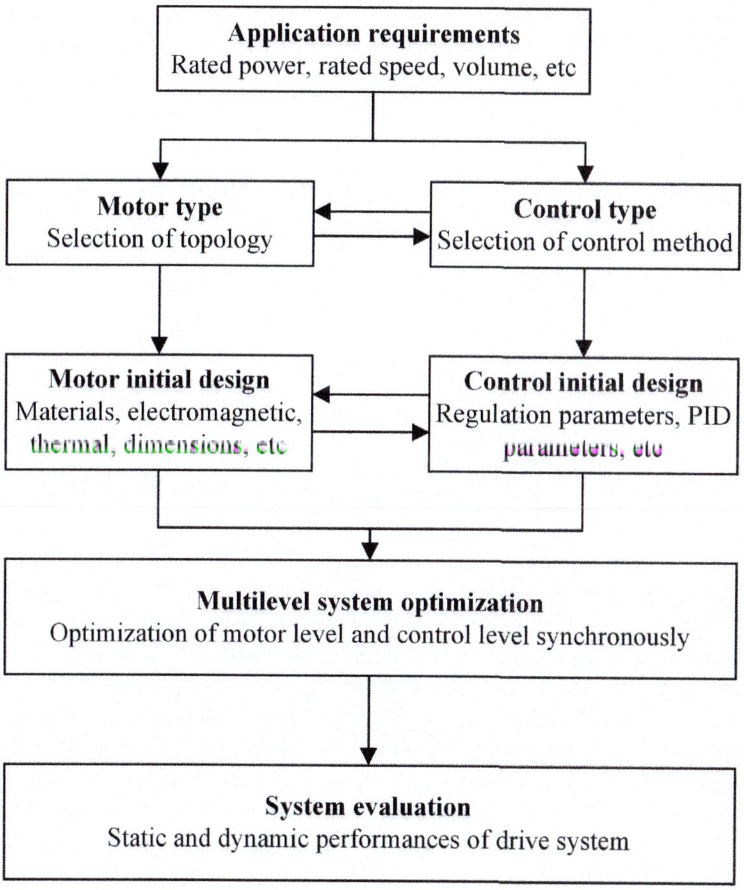

Fig. 6.1 System level design and optimization framework for electrical drive systems

In this stage, the coupled influences between multi-physics and the control aspects are taken into consideration at the same time. Many trade-offs exist among motor types for most industrial and domestic applications. All the advantages and drawbacks should be thoroughly considered based on the operation conditions, such as cooling capabilities, torque density, and the total cost. There are many applications where SRMs are considered as the best alternatives due to their low cost and superior thermal and mechanical robustness. Nowadays, the applications of SRMs in EV/HEVs, aircraft, and wind generators are the three main research aspects. The work presented in [28] proved that SRM exhibits better high-speed performance than the Prius motor. The SRMs designed for aerospace and wind power generation are presented in [29, 30], respectively.

Step 3: Accomplish the initial designs for both the motor and control types.

Most design approaches rely on the existing theories and the FEA method. The design process of SRMs is inherently empirical due to their high nonlinear characteristics, and no general mathematical model can accurately represent SRM performance. Safety margins of critical machine parameters, such as current density and cooling capability, and the slot fill coefficient are often employed to prevent overheating and assembly difficulty, which limits the designed machine to the ideal state.

Step 4: Optimize the models for the whole system, including the motor and control levels.

The optimization problem can be defined as [7]

$$\begin{aligned} &\min : f(\mathbf{x}_s) \\ &\text{s.t.} \quad g_i(\mathbf{x}_s) \leq 0, i = 1, 2, \ldots, n \\ &\qquad \mathbf{x}_{sl} \leq \mathbf{x}_s \leq \mathbf{x}_{su} \end{aligned} \tag{6.1}$$

$$\begin{aligned} &\min : f(\mathbf{x}_s) \\ &\text{s.t.} \quad g_i(\mathbf{x}_s) \leq 0, i = 1, 2, \ldots, n \\ &\qquad \mathbf{x}_{sl} \leq \mathbf{x}_s \leq \mathbf{x}_{su} \end{aligned} \tag{6.2}$$

where $\mathbf{x}_s, f$, and g_i are the design parameter vector, objective, and constraints of the motor, respectively, $\mathbf{x}_s$ consists of the motor parameter vector and control parameter vector, and $\mathbf{x}_{sl}$ and $\mathbf{x}_{su}$ are the lower and upper boundaries, respectively.

There are two strategies for optimizing the model (6.1): single space and sequential subspace strategies. The single space strategy simultaneously optimizes all motor and control parameters using an algorithm like GA and DEA. It requires a huge computation cost of FEA and control simulation as (6.1) is normally a high-dimensional optimization problem (dimension usually no less than 10). To improve the optimization efficiency, a sequential subspace optimization strategy was presented for electrical drive systems [7, 8].

Step 5: Evaluation of the system's performance.

This step can be divided into two aspects, i.e., the static performance of the motor and the dynamic performance of the whole drive system. The static performance of SRMs can reflect the characteristics of motor topology, and the drive system's dynamic performance can evaluate the motor and control integration.

6.3 Multi-level Optimization Method

Electrical machines, such as motors and generators, are complex systems that involve multiple design variables and objectives. The optimization of these systems is essential to achieve high performance and efficiency. The multi-level optimization method is a powerful approach that can efficiently optimize the electrical machines at different levels of design variables and objectives [31–33].

Figure 6.2 shows a flowchart of the proposed multi-objective system-level optimization method for SRM drive systems. This method includes sequential subspace and system-level optimizations and can be divided into seven steps.

Step 1: Build the multi-objective optimization problem by considering the system parameters for the SRM drive systems and determining the selection criterion to choose the optimal solution.

The multi-objective model can be defined as

$$\begin{aligned} \min: &\begin{cases} f_1(\mathbf{x}_\mathrm{s}) \\ f_2(\mathbf{x}_\mathrm{s}) \\ \vdots \\ f_m(\mathbf{x}_\mathrm{s}) \end{cases}. \\ \text{s.t.} \quad & g_i(\mathbf{x}_\mathrm{s}) \le 0, i = 1, 2, \ldots, n \\ & \mathbf{x}_\mathrm{sl} \le \mathbf{x}_\mathrm{s} \le \mathbf{x}_\mathrm{su} \end{aligned} \tag{6.3}$$

The selection criterion can be defined as

$$\min: \quad F\{f_1(\mathbf{x}_\mathrm{s}), f_2(\mathbf{x}_\mathrm{s}), \ldots, f_m(\mathbf{x}_\mathrm{s})\}. \tag{6.4}$$

Step 2: Carry out sensitivity analysis for all parameters of the SRM drive system, including motor level and control level.

Sensitivity analysis is suitable for high-dimensional optimization problems to divide all the parameters further into different levels. Optimizing all the parameters in one subspace is time-consuming, and common approximate methods lack sufficient accuracy in solving high-dimensional problems. This method contributes to reducing computing costs and building approximate models. Four types of sensitivity analysis methods are commonly used, i.e., local sensitivity analysis, global sensitivity analysis, sizing equation, and analysis of variance.

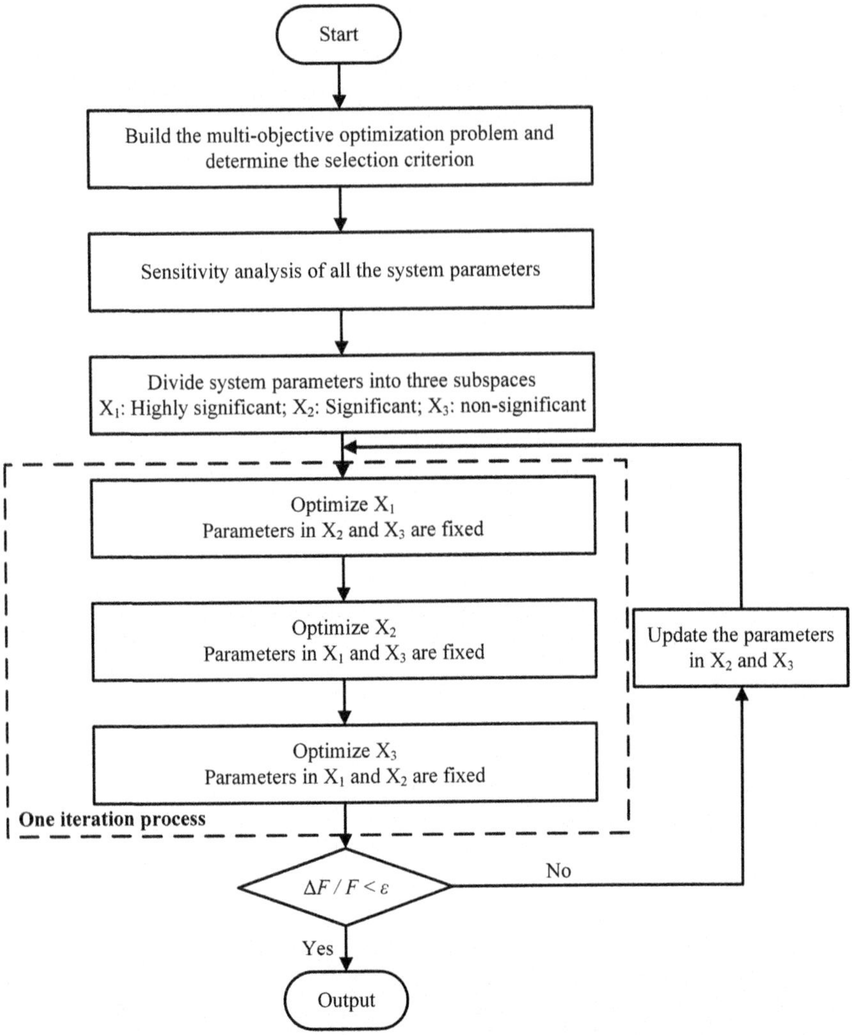

Fig. 6.2 A flowchart of multi-level optimization method for SRM drive systems

Step 3: Divide the system parameters into three subspaces X_1, X_2, and X_3, where X_1, X_2, and X_3 represent the highly significant, significant, and non-significant subspaces of system design parameters, respectively.

The results of influences on the defined optimization objectives decide the division's principle (6.3). The results can be achieved from the sensitivity analysis or the design experience.

Step 4: Optimize subspace X_1.

In the implementation, parameters in X_2 and X_3 are fixed as the initial values at the first iteration. The approximate models and intelligent algorithms are applied in the optimization process. Firstly, develop a parametric finite element model (FEM) in X_1 for the investigated SRM and implement the simulation by using a design of experiment (DoE) technique. Secondly, obtain the simulation results and evaluate the motor performance. Thirdly, build an approximate model based on the obtained results and motor performance. This approximate model can significantly reduce computational costs. Finally, an intelligent algorithm is employed to find the optimal solutions for the motor. The Pareto optimal solution set of X_1 is obtained in this step, and then the optimal solution is selected for the optimization of the next subspace X_2 according to (6.4).

Step 5: Optimize subspace X_2.

Similar to the last step, the Pareto optimal solution set of X_2 is obtained in this step based on the selected values of parameters in X_1. It should be noted that in this process, the parametric FEM is related to the variables in X_2. After developing the optimization problem and implementing the optimization algorithm, the optimal solutions of X_2, together with those of X_1 obtained in step 4, will be sent to optimize the next subspace X_3.

Step 6: Optimize subspace X_3.

The values of parameters in X_1 and X_2 are fixed as selected in this step. Similarly, after the development of FEM and the optimization problem, the Pareto optimal solution set of X_3 can be achieved by using an optimization algorithm.

It should be noted that the whole optimization of X_1, X_2, and X_3 is regarded as one iteration process. After the optimization of subspace X_3, the optimal solution of this iteration can be achieved.

Step 7: Termination judgment.

Compare the objective functions between two successive iterations. If their relative error is smaller than a given value, terminate the iteration progress and output the optimization results/Pareto front. Otherwise, update the parameters in X_2 and X_3 and go to step 4 for another iteration. Steps 4–7 can be regarded as a sequential optimization process.

Applying the multi-level optimization method can lead to significant improvements in the performance and efficiency of electrical machines. It can optimize the design variables at different levels, leading to a balanced solution that satisfies the requirements of multiple objectives. Furthermore, the method can reduce the computational cost and time required for the optimization process.

6.4 Design Example of System-Level Deterministic Optimization

6.4.1 *Multi-objective Optimization Targets*

In this example, an SRSRM is designed for belt-driven starter/generator (BSG) applications in hybrid electric vehicles (HEVs). The specifications of SRSRM are listed in Table 6.1. An SRM drive system, comprised of this SRSRM and an angular position control system, will be investigated.

Figure 6.3 shows the topology of the SRSRM. As shown, the stator of this SRSRM is composed of excited and auxiliary poles. While the excited poles are wound by windings, the auxiliary poles only function as flux return paths with no windings. The rotor contains a series of discrete segmented rotors, and each component is embedded in the nonmagnetic isolator [34]. Aside from the number of phases, stator, and rotor, the other ten parameters listed in Table 6.2 are considered the optimization parameters, including the parameters of motor level and control level. It should be noted that the diameter of the winding is kept constant during the optimization process. The main reason for this assumption is that it can simplify the optimization since it is hard to reflect the change of the winding's diameter for the SRM in FEA.

Meanwhile, in the practical design process of an SRM, there are several feasible combinations of the winding diameter and the number of turns for a specific slot fill factor. A common practice is to select the diameter first, then determine the optimal number of turns according to the slot fill factor and other constraints. Furthermore,

Table 6.1 Specifications of the SRSRM

Parameters	Unit	Value
Rated power	kW	1.8
Rated speed	r/min	6000
Rated voltage	V	60
Efficiency	%	85
Outer diameter	mm	128
Axial length	mm	80

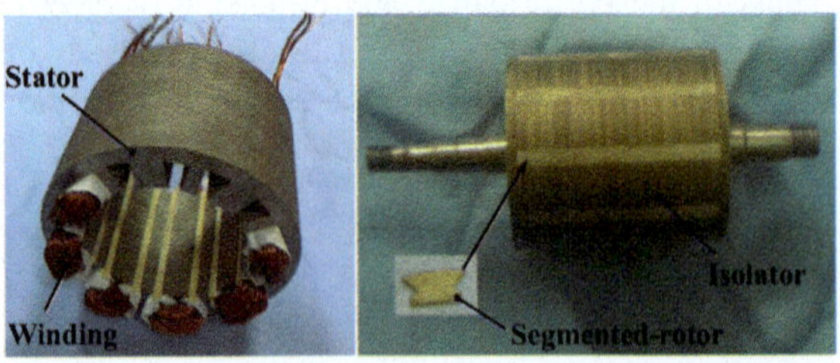

Fig. 6.3 Topology of the SRSRM

Table 6.2 Initial design values of the SRSRM

Par.	Description	Unit	Value
m	Number of phases	–	4
N_s	Number of stator poles	–	16
N_r	Number of rotor poles	–	10
D_{so}	Stator outer diameter	mm	128
l	Axial length	mm	80
D	Rotor outer diameter	mm	82
β_{s1}	Excited stator pole arc	°	21.38
β_{s2}	Auxiliary stator pole arc	°	10.69
β_r	Rotor pole arc	°	26.64
L_{sy}	Stator yoke	mm	7
h_{cr}	Height of segmented rotor	mm	5.5
g	Air gap	mm	0.25
N	Number of turns	–	24
θ_{on}	Turn-on angle	°	– 3
θ_{off}	Turn-off angle	°	12

the influence of the winding diameter on the resistance can also be reflected by the number of turns for a given slot fill factor.

For this SRSRM drive system, in addition to the torque and loss, the torque ripple is another important characteristic compared to the other types of machines, as it is the main obstacle for the application of SRMs in industry. The loss is the sum of the copper loss and iron loss. In this application, the optimization problem can be defined as follows:

$$
\begin{aligned}
\min: & \begin{cases} f_1(\mathbf{x}_s) = -T_{avg} \\ f_2(\mathbf{x}_s) = P_{loss} \\ f_3(\mathbf{x}_s) = T_{ripple} \end{cases} \\
\text{s.t.} \quad & g_1(\mathbf{x}_s) = 0.85 - \eta \leq 0 \\
& g_2(\mathbf{x}_s) = sf - 0.6 \leq 0 \\
& g_3(\mathbf{x}_s) = J_c - 6 \leq 0 \\
& \mathbf{x}_{sl} \leq \mathbf{x}_s \leq \mathbf{x}_{su}
\end{aligned}
\tag{6.5}
$$

where T_{avg}, P_{loss}, and T_{ripple} are the three optimization objectives, which represent the average output torque, average loss, and torque ripple, respectively; η, sf, and J_c represent the efficiency, slot fill factor, and current density, respectively. It should be noted that although $f_1(\mathbf{x}_s)$ and $f_2(\mathbf{x}_s)$ can reflect the efficiency, they do not directly show whether the efficiency meets the requirement. The efficiency specifications are listed in Table 6.1. Thus, it has been taken as one constraint.

To select the optimal solution for a practical application, the selection criterion is defined as

$$\min: \quad F = w_1 \frac{T_{\text{avg\_initial}}}{T_{\text{avg}}} + w_2 \frac{P_{\text{loss}}}{P_{\text{loss\_initial}}} + w_3 \frac{T_{\text{ripple}}}{T_{\text{ripple\_initial}}} \tag{6.6}$$

where $T_{\text{avg\_initial}}$, $P_{\text{loss\_initial}}$, and $T_{\text{ripple\_initial}}$ are the average torque, loss, and torque ripple of the initial design, and w_1, w_2, and w_3 are the weight factors. In this example, w_1, w_2, and w_3 are assigned 0.4, 0.3, and 0.3, respectively. The single objective function (6.6) is adopted for the determination of the convergence of the Pareto front since it can provide a fast optimization process [7].

6.4.2 Sensitivity Analysis

The local sensitivity analysis [34, 35], as presented in Chap. 5, is performed to divide all the design parameters into three subspaces.

The sensitivity analysis is based on the FEA, as there is no accurate analytical expression of the objective function. Thus, a differential form is taken to calculate S_i by

$$S_i = \frac{f(\mathbf{x}_0 \pm \Delta \mathbf{x}_i) - f(\mathbf{x}_0)}{\pm \Delta x_i}. \tag{6.7}$$

It should be noted that the sensitivities of different parameters calculated by (6.7) have different units. Thus, a normalization step considering absolute value is carried out as follows:

$$SA_i = \left| \frac{\left[f(\mathbf{x}_0 \pm \Delta \mathbf{x}_i) - f(\mathbf{x}_0)\right] / f(\mathbf{x}_0)}{\pm \Delta x_i / x_i} \right|. \tag{6.8}$$

In general, there are two methods to determine the increment Δx_i. The first is the parameter variation method, in which Δx_i is usually defined as 10, 20%, or both of their initial values. The other is the deviation variation method, in which Δx_i is generally defined as the standard deviation of x_i. This paper will investigate the first method as the deviation is not given. It should be noted that the initial absolute value and variation range of θ_{off} are both four times those of θ_{on}, respectively. Thus, the local analysis of θ_{on} and θ_{off} can be referenced in this example.

The space requirement of the application determines the stator's outer diameter and the axial length. Excluding m, N_s, N_r, D_{so}, and l, the other ten parameters listed in Table 6.2 are considered optimization variables. The calculation requires 41 samples, including 40 for those four variations (− 20, − 10, 10, and 20%) and one initial sample. The four relative errors of each objective function are calculated by (6.8), and then the average of absolute values of these four values represents the sensitivity of the specific objective function.

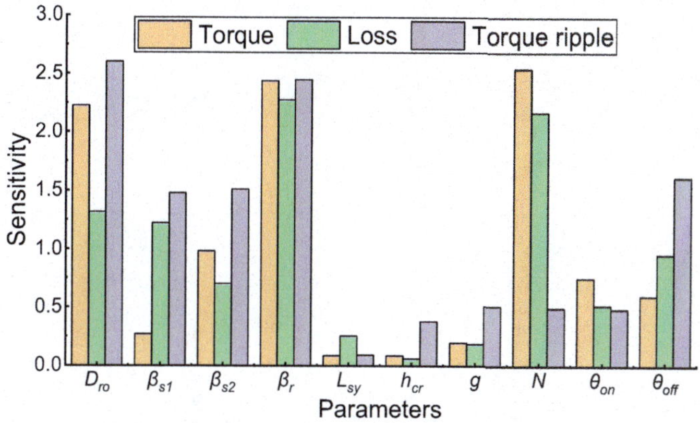

Fig. 6.4 Local sensitivity indices of torque, loss, and torque ripple

Figure 6.4 shows the local sensitivity analysis results of the ten parameters. Three subspaces are determined according to the sensitivity influence on the defined optimization objectives, i.e., average torque, loss, and torque ripple. Subspace X_1 includes parameters D, β_r, and N; subspace X_2 includes parameters β_{s1}, β_{s2}, and θ_{off}, and subspace X_3 includes parameters θ_{on}, L_{sy}, h_{cr}, and g.

6.4.3 Multi-level Optimization Process

In the process of multi-objective optimization, the Kriging model is used to approximate the SRM drive model, and NSGA II is selected as the optimization algorithm. For this SRM drive system, FEM is employed to calculate the torque, loss, and torque ripple under the APC method at the rated speed of 6000 rev/min. As the speed increases, the value of torque ripple decreases. Since the torque ripple of SRM is directly related to the structure due to its double salient poles, the change of torque ripple has the same tendency under the whole speed range. Therefore, the value of torque ripple under fixed speed can reflect this performance of the entire speed range.

As illustrated in Fig. 6.2, the optimization will be terminated when the result of F defined in (6.6) meets the convergence criterion. In total, three iteration processes are carried out in this example, where each iteration contains the sequential optimization of the three subspaces of X_1, X_2, and X_3. Figures 6.5, 6.6 and 6.7 show the Pareto optimal solutions of each subspace in 3 iterations.

The parameters in subspace X_1 are highly significant to the overall performances of the drive system, and they are optimized at the first subspace. In this subspace, 1001 ($11 \times 13 \times 7$, where 11, 13, and 7 are the sampling numbers of parameters D_{ro}, β_r, and N, respectively) FEM samples are simulated for the development of the Kriging model. After the optimization of subspace X_1, significant parameters in

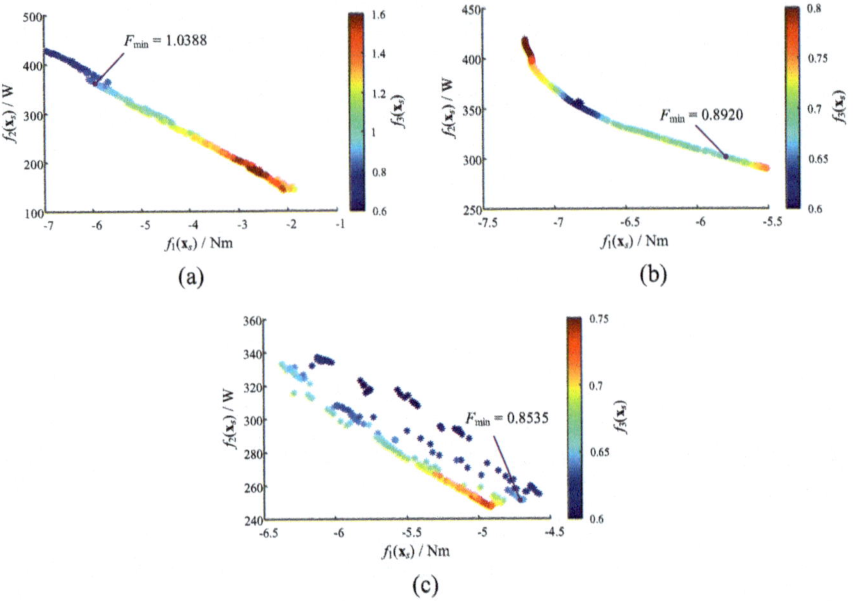

Fig. 6.5 Pareto optimal solutions of iteration 1. **a** Subspace X_1, **b** subspace X_2, and **c** subspace X_3

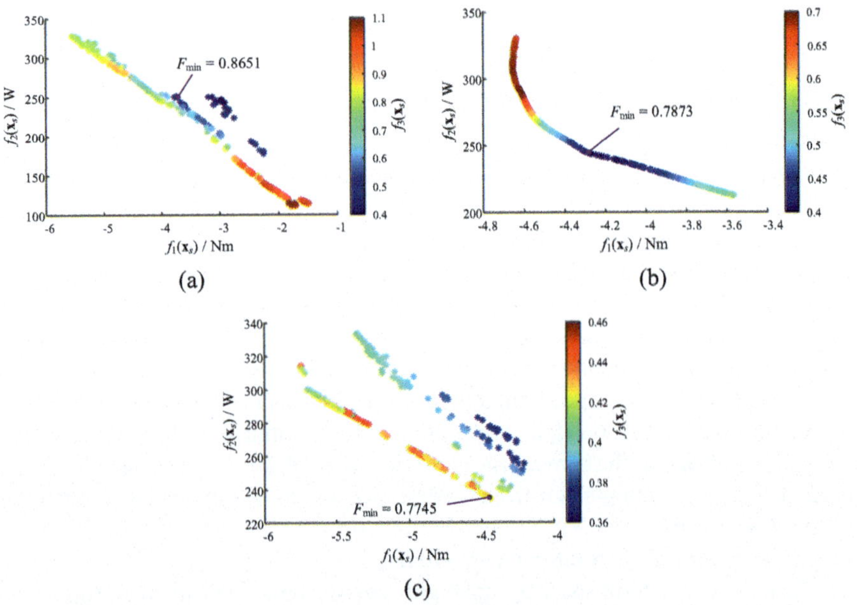

Fig. 6.6 Pareto optimal solutions of iteration 2. **a** Subspace X_1, **b** subspace X_2, and **c** subspace X_3

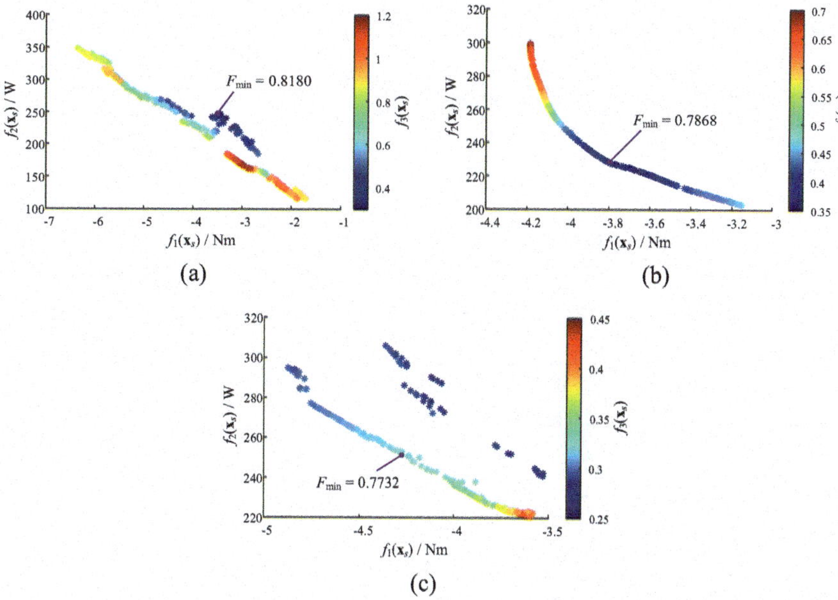

Fig. 6.7 Pareto optimal solutions of iteration 3. **a** Subspace X_1, **b** subspace X_2, and **c** subspace X_3

subspace X_2 will be optimized with the selected parameters in X_1. For the optimal solution in X_2, 150 ($6 \times 5 \times 5$, where 6, 5, and 5 mean the sampling numbers of parameters β_{s1}, β_{s2}, and θ_{off}, respectively) FEM samples are sampled, and the corresponding Pareto solutions can be achieved during the optimization of subspace X_2. Similarly, for the optimal solution in X_3, 180 ($3 \times 3 \times 5 \times 4$, where 3, 3, 5, and 4 mean the numbers of values of θ_{on}, L_{sy}, h_{cr}, and g, respectively) FEM samples are required for the optimization of subspace X_3.

From Fig. 6.5a, it can be found that the proposed method can provide a set of optimal solutions for situations with different output power demands. The solution with a minimum value of F defined in (6.6) is sent to the optimization of subspace X_2. In the same way, an optimal solution with a minimum value of F is sent to the next subspace. Comparing Fig. 6.5a, b, the torque ripple has been reduced significantly. This means that the optimization of subspace X_2 can help reduce the torque ripple of the SSRM. Table 6.3 lists the optimal solutions of Iterations 1, 2, and 3. The value of ε (as illustrated in Fig. 6.2) is set as 1%. Since the value of $\Delta F/F$ of Iterations 2 and 3 is 0.17%, less than 1%, the optimization is terminated after three iterations.

Thus, the final optimal solution is the optimal result of Iteration 3. The value of F gradually decreases as the iteration number increases, as shown in Figs. 6.5c, 6.6c and 6.7c, and Table 6.3.

On the other hand, sequential subspace optimization can greatly improve the computational efficiency of the whole system optimization. The direct one-level optimization, for example, optimizing all ten parameters by using FEM and NSGA

Table 6.3 Optimization results

Par.	Unit	Initial	Iteration 1	Iteration 2	Iteration 3
D_{ro}	mm	82.00	84.98	84.66	77.11
β_{s1}	°	21.38	21.50	21.50	21.50
β_{s2}	°	10.69	10.45	9.00	9.02
β_r	°	26.64	25.46	27.00	27.04
L_{sy}	mm	7.00	8.56	8.94	9.00
h_{cr}	mm	5.5	6.95	5.82	5.97
g	mm	0.25	0.3	0.25	0.25
N	–	24	20	20	20
θ_{on}	°	− 3.00	− 2.50	− 2.50	− 2.85
θ_{off}	°	12.00	11.00	11.00	11.00
T_{avg}	N m	3.03	4.72	4.43	4.30
P_{out}	kW	1.90	2.97	2.78	2.70
P_{loss}	W	195.06	250.68	235.18	252.94
η	%	90.69	92.22	92.20	91.43
T_{ripple}	%	93.64	65.03	43.44	31.94
F	–	1.00	0.8507	0.7745	0.7732

II, requires around 50,000 FEM samples (500 × 100, where 500 is the average iteration number of NSGA II and 100 is the population size in each iteration) are required. If the Kriging model is employed, the required FEM samples (1001 × 150 × 180) will be much larger. Regarding the sequential subspace optimization, if one optimal point is selected, only 1331 (1001 + 150 + 180) FEM samples are required for one iteration in the optimization of this drive system. During the optimization process, three iterations are performed, and a total of 3993 FEM samples are required. Therefore, the proposed sequential subspace optimization method can significantly reduce the burden and cost of computation, especially for the optimization of high-dimensional structures. Besides, the computation time of each subspace is listed in Table 6.4, where T_FEM, T_Kriging, and T_NSGA II mean the computer time of FEM, Kriging model, and NSGA II, respectively. The FEM almost consumes the computation time, while the Kriging model and NSGA II only take up several seconds. The computation time of each iteration is the sum of the three subspaces, and the final total time is the product of the iteration number and the optimization time of each iteration.

Table 6.4 Computation time of each subspace

Subspace	T_FEM (h)	T_Kriging (s)	T_NSGA II (s)
X_1	25	6.49	13.10
X_2	3.73	0.11	2.39
X_3	4.5	0.26	2.55

Table 6.5 Comparison between Kriging model and FEM

Par.	Iteration 1		Iteration 2		Iteration 3	
	Krig.	FEM	Krig.	FEM	Krig.	FEM
T_{avg} (N m)	4.72	4.68	4.43	4.31	4.30	4.22
P_{loss} (W)	250.68	254.77	235.18	244.57	252.94	244.32
T_{ripple} (%)	65.03	63.06	43.44	42.92	31.94	32.88
T_{avg} error (%)	0.85		2.71		1.86	
P_{loss} error (%)	1.63		3.99		0.35	
T_{ripple} error (%)	3.03		1.20		1.86	

6.4.4 *Verification of the Kriging Model*

The Kriging model is employed to reduce the burden of computation. To ensure the optimization efficiency, it is necessary to verify the accuracy of the Kriging model. After the optimization process, the optimal solutions for each iteration are selected. Thus, the results between the Kriging model and FEM of these points are compared to verify the effectiveness of the approximate model. The FEM results are verified by comparing with the experimental results, as mentioned above. Thus, the accuracy of the Kriging model can be verified by the FEM results.

Table 6.5 lists the values of the three optimization objectives calculated by the Kriging model and FEM. As shown, the maximum errors of these three optimal points' torque, loss, and torque ripple are only 3.03%, 3.99%, and 1.86%, respectively. Thus, the accuracy of the Kriging model is acceptable.

6.4.5 *Result Comparison*

After verification of the Kriging model, the optimal solutions shown in Figs. 6.5, 6.6 and 6.7 are credible. The comparisons of torque and current between the optimal solution and initial design are illustrated in Figs. 6.8 and 6.9, respectively. As shown in Fig. 6.8, the SSRM with the selected optimal design has a higher average output torque and a smaller torque ripple than the motor with the initial design. As shown in Fig. 6.9, compared with the initial design, the waveform of the optimal design is closer to the ideal flat waveform. Table 6.3 lists the detailed comparison of values between the initial design and the final selected optimal solution (optimum in the third iteration).

The initial design has an average torque of 3.03 N m, an efficiency of 90.69%, and a torque ripple of 93.64%. After optimizing the whole system using the proposed method, the motor can achieve an average torque of 4.30 N m (an increase of about 41.9%) with a torque ripple of only 31.94% (a reduction of about 65.89%). Therefore,

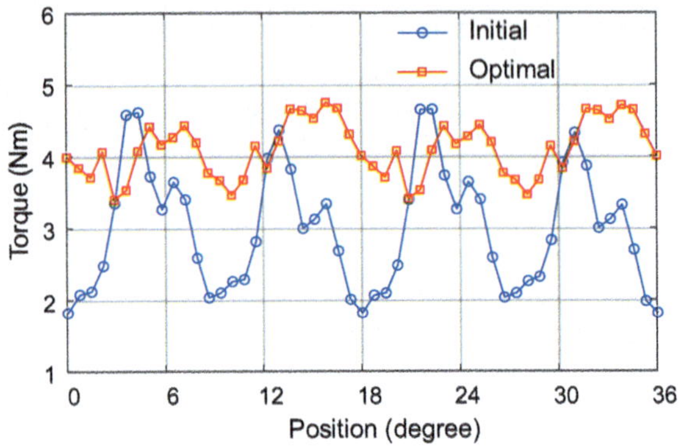

Fig. 6.8 Comparison of torque between the initial and optimal solutions

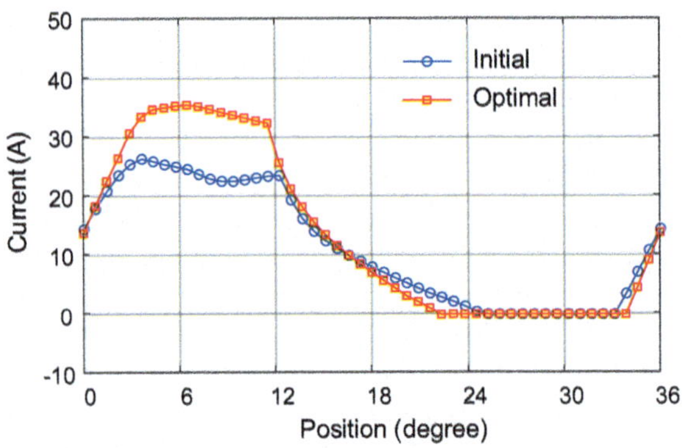

Fig. 6.9 Comparison of current between the initial and optimal solutions

the proposed method can greatly reduce the problem of high torque ripples in an SSRM drive without sacrificing its average torque.

6.5 Summary

This chapter presents a novel multi-objective and system-level design optimization method for SRM drive systems. The proposed method aims to provide a fast way to achieve the best performance of the SRM drive system. A drive system comprised of an SRSRM and APC method is investigated to demonstrate the effectiveness of the proposed method. To improve the system's overall performance and reduce the computation cost, all the parameters at the motor and control levels are considered together and divided into three optimization subspaces using the sensitivity analysis results. The optimal solution obtained by the proposed method exhibits many benefits for the drive system, including high efficiency and low torque ripple.

References

1. Wang D, Du X, Zhang D, Wang X. Design, optimization, and prototyping of segmental-type linear switched-reluctance motor with a toroidally wound mover for vertical propulsion application. IEEE Trans Ind Electron. 2018;65(2):1865–74. https://doi.org/10.1109/TIE.2017.2740824.
2. Mousavi-Aghdam SR, Feyzi MR, Bianchi N, Morandin M. Design and analysis of a novel high-torque stator-segmented SRM. IEEE Trans Ind Electron. 2016;63(3):1458–66. https://doi.org/10.1109/TIE.2015.2494531.
3. Asgar M, Afjei E, Torkaman H. A new strategy for design and analysis of a double-stator switched reluctance motor: electromagnetics, FEM, and experiment. IEEE Trans Magn. 2015;51(12):1–8. https://doi.org/10.1109/TMAG.2015.2465307.
4. Ma J, Li J, Fang H, Li Z, Liang Z, Fu Z, et al. Optimal design of an axial-flux switched reluctance motor with grain-oriented electrical steel. IEEE Trans Ind Appl. 2017;53(6):5327–37. https://doi.org/10.1109/TIA.2017.2727438.
5. Sun X, Shi Z, Lei G, Guo Y, Zhu J. Analysis and design optimization of a permanent magnet synchronous motor for a campus patrol electric vehicle. IEEE Trans Veh Technol. 2019;68(11):10535–44. https://doi.org/10.1109/TVT.2019.2939794.
6. Diao K, Sun X, Bramerdorfer G, Cai Y, Lei G, Chen L. Design optimization of switched reluctance machines for performance and reliability enhancements: a review. Renew Sust Energ Rev. 2022;168. https://doi.org/10.1016/j.rser.2022.112785.
7. Lei G, Wang T, Guo Y, Zhu J, Wang S. System-level design optimization methods for electrical drive systems: deterministic approach. IEEE Trans Ind Electron. 2014;61(12):6591–602. https://doi.org/10.1109/TIE.2014.2321338.
8. Lei G, Wang T, Zhu J, Guo Y, Wang S. System-level design optimization method for electrical drive systems—robust approach. IEEE Trans Ind Electron. 2015;62(8):4702–13. https://doi.org/10.1109/TIE.2015.2404305.
9. Diao K, Sun X, Lei G, Guo Y, Zhu J. Multiobjective system level optimization method for switched reluctance motor drive systems using finite-element model. IEEE Trans Ind Electron. 2020;67(12):10055–64. https://doi.org/10.1109/tie.2019.2962483.
10. Bilgin B, Emadi A, Krishnamurthy M. Design considerations for switched reluctance machines with a higher number of rotor poles. IEEE Trans Ind Electron. 2012;59(10):3745–56. https://doi.org/10.1109/TIE.2011.2141102.
11. Sun X, Diao K, Yang Z. Performance improvement of a switched reluctance machine with segmental rotors for hybrid electric vehicles. Comput Electr Eng. 2019;77:244–59. https://doi.org/10.1016/j.compeleceng.2019.06.002.
12. Sun X, Diao K, Yang Z, Lei G, Guo Y, Zhu J. Direct torque control based on a fast modeling method for a segmented-rotor switched reluctance motor in HEV application. IEEE J Emerg Sel Top Power Electron. 2019;1. https://doi.org/10.1109/JESTPE.2019.2950085.
13. Fleming FE, Edrington CS. Real-time emulation of switched reluctance machines via magnetic equivalent circuits. IEEE Trans Ind Electron. 2016;63(6):3366–76. https://doi.org/10.1109/TIE.2016.2521343.
14. Nabeta SI, Chabu IE, Lebensztajn L, Correa DAP, Silva WMd, Hameyer K. Mitigation of the torque ripple of a switched reluctance motor through a multiobjective optimization. IEEE Trans Magn. 2008;44(6):1018–21. https://doi.org/10.1109/TMAG.2007.915137.
15. Lee JH. Optimum shape design solution of flux switching motor using response surface methodology and new type winding. IEEE Trans Magn. 2012;48(4):1637–40. https://doi.org/10.1109/TMAG.2011.2173564.
16. Cai J, Deng ZQ, Qi RY, Liu ZY, Cai YH. A novel BVC-RBF neural network based system simulation model for switched reluctance motor. IEEE Trans Magn. 2011;47(4):830–8. https://doi.org/10.1109/TMAG.2011.2105273.
17. Sun X, Diao K, Lei G, Guo Y, Zhu J. Real-time HIL emulation for a segmented-rotor switched reluctance motor using a new magnetic equivalent circuit. IEEE Trans Power Electron. 2019;1. https://doi.org/10.1109/TPEL.2019.2933664.

18. Kano Y, Kosaka T, Matsui N. Optimum design approach for a two-phase switched reluctance compressor drive. IEEE Trans Ind Appl. 2010;46(3):955–64. https://doi.org/10.1109/TIA.2010.2045212.
19. Zhang J, Wang H, Chen L, Tan C, Wang Y. Multi-objective optimal design of bearingless switched reluctance motor based on multi-objective genetic particle swarm optimizer. IEEE Trans Magn. 2018;54(1):1–13. https://doi.org/10.1109/TMAG.2017.2751546.
20. Rallabandi V, Wu J, Zhou P, Dorrell DG, Ionel DM. Optimal design of a switched reluctance motor with magnetically disconnected rotor modules using a design of experiments differential evolution FEA-based method. IEEE Trans Magn. 2018;54(11):1–5. https://doi.org/10.1109/TMAG.2018.2850744.
21. Deng, Mecrow B, Wu H, Martin R, Gai Y. Cost-effective and high-efficiency variable-speed switched reluctance drives with ring-connected winding configuration. IEEE Trans Energy Convers. 2019;34(1):120–9. https://doi.org/10.1109/TEC.2018.2851065.
22. Ding W, Yang S, Hu Y. Development and investigation on segmented-stator hybrid-excitation switched reluctance machines with different rotor pole numbers. IEEE Trans Ind Electron. 2018;65(5):3784–94. https://doi.org/10.1109/TIE.2017.2760846.
23. Feng L, Sun X, Tian X, Diao K. Direct torque control with variable flux for an SRM based on hybrid optimization algorithm. IEEE Trans Power Electron. 2022;37(6):6688–97. https://doi.org/10.1109/tpel.2022.3145873.
24. Li X, Shamsi P. Model predictive current control of switched reluctance motors with inductance auto-calibration. IEEE Trans Ind Electron. 2016;63(6):3934–41. https://doi.org/10.1109/TIE.2015.2497301.
25. Feng L, Sun X, Bramerdorfer G, Zhu Z, Cai Y, Diao K, et al. A review on control techniques of switched reluctance motors for performance improvement. Renew Sust Energ Rev. 2024;199:114454. https://doi.org/10.1016/j.rser.2024.114454.
26. Diao K, Bramerdorfer G, Sun X, Yang Z, Han S. Multiobjective design optimization of a novel switched reluctance motor with unequal alternating stator yoke segments. IEEE Trans Transp Electrification. 2023;9(1):512–21. https://doi.org/10.1109/tte.2022.3174337.
27. Diao K, Sun X, Lei G, Bramerdorfer G, Guo Y, Zhu J. System-level robust design optimization of a switched reluctance motor drive system considering multiple driving cycles. IEEE Trans Energy Convers. 2021;36(1):348–57. https://doi.org/10.1109/tec.2020.3009408.
28. Kiyota K, Kakishima T, Chiba A. Comparison of test result and design stage prediction of switched reluctance motor competitive with 60-kW rare-earth PM motor. IEEE Trans Ind Electron. 2014;61(10):5712–21. https://doi.org/10.1109/TIE.2014.2304705.
29. Bartolo JB, Degano M, Espina J, Gerada C. Design and initial testing of a high-speed 45-kW switched reluctance drive for aerospace application. IEEE Trans Ind Electron. 2017;64(2):988–97. https://doi.org/10.1109/TIE.2016.2618342.
30. Liu X, Park K, Chen Z. A novel excitation assistance switched reluctance wind power generator. IEEE Trans Magn. 2014;50(11):1–4. https://doi.org/10.1109/TMAG.2014.2327798.
31. Diao K, Sun X, Lei G, Guo Y, Zhu J. Application-oriented system level optimization method for switched reluctance motor drive systems. In: 2020 IEEE 9th International power electronics and motion control conference (IPEMC2020-ECCE Asia). 2020. p. 472–7.
32. Sun X, Shi Z, Lei G, Guo Y, Zhu J. Multi-objective design optimization of an IPMSM based on multilevel strategy. IEEE Trans Ind Electron. 2021;68(1):139–48. https://doi.org/10.1109/TIE.2020.2965463.
33. Lei G, Liu C, Zhu J, Guo Y. Techniques for multilevel design optimization of permanent magnet motors. IEEE Trans Energy Convers. 2015;30(4):1574–84. https://doi.org/10.1109/tec.2015.2444434.
34. Sun X, Diao K, Lei G, Guo Y, Zhu J. Study on segmented-rotor switched reluctance motors with different rotor pole numbers for BSG system of hybrid electric vehicles. IEEE Trans Veh Technol. 2019;68(6):5537–47. https://doi.org/10.1109/TVT.2019.2913279.
35. Morio J. Global and local sensitivity analysis methods for a physical system. Eur J Phys. 2011;32(6):1577–83. https://doi.org/10.1088/0143-0807/32/6/011.

Chapter 7
Multi-objective Robust Optimization Method for Switched Reluctance Motor Drive Systems

Abstract In the last chapter, the deterministic optimization method was proposed for switched reluctance motor (SRM) drive systems without considering parameter uncertainties in production. Many unavoidable uncertainties exist in the practical production of SRM drive systems, mainly caused by manufacturing tolerances. These uncertainties/tolerances will result in significant variations in the practical performance of the SRMs, affecting the reliability and quality of motors and drive systems in mass production. This chapter presents a robust design optimization approach based on the sequential Taguchi method for SRM drive systems. The component-level and system-level methods are implemented and compared. Besides, deterministic and robust approaches are investigated using a design example. The comparison shows that the proposed robust optimization method can significantly improve the robustness of the investigated SRM drive system.

Keywords Multi-objective optimization · Robust optimization · Switched reluctance motors · Taguchi method · Manufacturing tolerance

7.1 Introduction

Previous studies have incorporated the deterministic approach with multi-objective, multi-level, multi-disciplinary, or multi-mode optimization methods [1–4]. For instance, in [5], a new multi-objective optimization method was proposed for a switched reluctance motor (SRM), in which six objectives regarding the requirements are specified. Furthermore, a system-level optimization method with multi-objective and multi-level optimization techniques was proposed in [6]. The sequential subspace optimization method and the surrogate model considered the motor and control parameters. The deterministic approach does not investigate the unavoidable uncertainties/noise in electrical machines' manufacturing or assembling stages. To reduce the computational cost, a multi-level design optimization method was

X. Sun et al., *Multi-objective Design Optimization of Switched Reluctance Motor Drive Systems*, https://doi.org/10.1007/978-981-96-0672-6_7

proposed [7], in which the optimization variables were divided into three levels, and each level was optimized independently. Moreover, in [8], a multi-mode optimization method was proposed for EV applications, and this optimization considers different motor operation modes based on the driving cycles.

The design optimization of SRMs is a multi-disciplinary, multi-objective, and highly nonlinear problem with unavoidable variables in manufacturing, which brings challenges to the industry [9–12]. The performance of the motor and drive system largely depends on the manufacturing accuracy. The manufacturing accuracy is determined by mechanical technology or processing conditions, such as the effects of punching or laser cutting [1, 13]. Changes in the SRM manufacturing process can cause uncertain fluctuations in output torque waveforms. Compared with the deterministic approach, design optimization considering manufacturing tolerances and variable environmental conditions for mass production is more complicated, as it includes not only the theoretical multi-disciplinary analysis but also the actual manufacturing experience of motors [14, 15].

Due to manufacturing tolerances and material diversity, variations of the design variables are unavoidable when producing SRMs. Mass production of SRMs with higher quality is a key requirement for decreasing the possibility of failure [6, 16, 17]. The six-sigma robust optimization method can effectively deal with optimization problems featuring unavoidable uncertainties in industrial manufacturing [18]. In [19], a robust design optimization approach was presented by incorporating the design for six sigma methodology, and it was verified by a case study of a transverse flux machine. The reliability and quality levels of the investigated motor were significantly improved. However, it requires abundant FEA-based evaluations, which leads to high computational costs. Compared with the six-sigma robust optimization method, the Taguchi design approach based on an orthogonal array exhibits high computational efficiency since it does not use complicated algorithms and additional programming [20, 21]. The Taguchi approach has successfully optimized different motors, such as interior permanent magnet synchronous motors [22] and SRMs [23].

The main drawback of the conventional Taguchi method is that the selection of the values and the improvement of performance significantly depend on the defined levels, which are highly dependent on the number of levels and human experience. Moreover, the previous work on the Taguchi method only considered the parameters for the motor domain but ignored the controller parameters, which can be regarded as a component-level Taguchi method rather than a system-level Taguchi method.

To solve the drawback of the conventional Taguchi method and achieve the robust optimal solution for the whole drive system, an efficient robust design optimization method will be presented for SRM drive systems in this chapter. A novel system-level sequential Taguchi method (SLSTM) is proposed to achieve a robust design for the whole SRM drive system, including the parameters in motor and control aspects [11]. The proposed method can improve the reliability and quality of products in batch production.

The remainder of this chapter is organized as follows. Section 7.2 presents the proposed SLSTM. Two design examples are investigated in Sects. 7.3 and 7.4, respectively, where the former compares the component-level and system-level robust optimizations, and the latter compares the system-level deterministic and robust optimization. Section 7.5 presents the summary.

7.2 System-Level Taguchi Parameter Design Method

The Taguchi parameter design method, also known as the Taguchi method, is a statistical approach that aims to optimize product or process performance by reducing the variability caused by the input parameters and noise factors.

In the context of electrical machines, the Taguchi method has been applied to improve efficiency, reduce losses, and enhance the performance of motors and generators. Taguchi method consists of three stages: design of experiments (DoE), signal-to-noise (S/N) ratio analysis, and optimization [24]. DoE involves selecting and varying the input parameters within their operating range and conducting limited experiments. The S/N ratio analysis quantifies the effect of the input parameters on the output performance and identifies the optimal combination of parameters. Optimization involves adjusting the parameter levels to achieve the desired output performance while considering the variability of the input parameters.

Figure 7.1 presents an optimization flowchart of the proposed SLSTM. It can be divided into the following six steps [11].

Step 1: Determine the system-level optimization model.

For a design optimization problem, the optimization model should be determined first. Multiple performance indicators, like the torque and torque ripple, can be included as the optimization objectives and/or constraints.

Step 2: Determine the levels for control and noise factors.

The control factors will be selected according to the sensitivity analysis of design parameters on each optimization objective. It should be noted that the design parameters for the proposed SLSTM contain both motor and control parameters, which jointly allow for system-level optimization.

Step 3: Generate the orthogonal array and implement the simulations.

An orthogonal array is a mathematical construct used in experimental design to efficiently and systematically study the effects of multiple factors on a process or system. In an orthogonal array, the factors being studied vary systematically across a set of experimental runs, such that each factor varies independently. This allows researchers to isolate and measure the effect of each factor on the process or system, while minimizing the number of experimental runs required. In this step, corresponding finite element samples are established and performed according to each row of the orthogonal array.

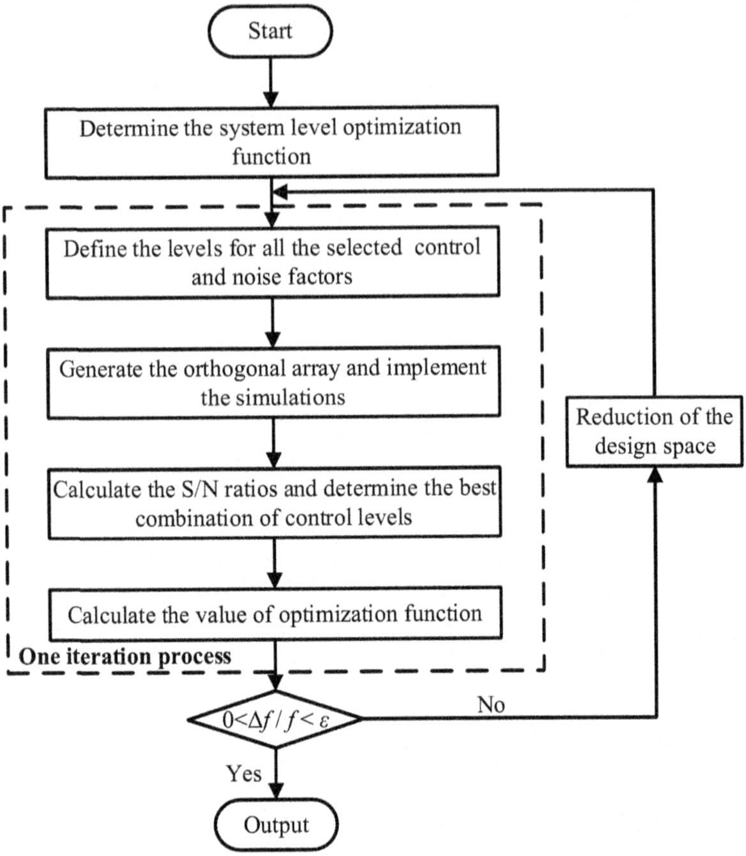

Fig. 7.1 A flowchart of the SLSTM

Step 4: Compute the S/N ratio and select the best value of each control factor.

According to the data obtained from the samples, the S/N ratio can be computed to find the optimal combination of control factors' levels. Two main steps should be accomplished to compute the S/N ratios. First, the S/N ratio for each row of the orthogonal array is calculated. Then, the average S/N ratios for all levels of optimization variables based on the S/N ratios of each row are computed.

Step 5: Calculate the value of the optimization function.

The value of the optimization function for the combined values of each parameter at the present iteration is calculated in this step. Steps 2–5 are considered one iteration process, distinguishing between the proposed SLSTM and the conventional Taguchi method.

Step 6: Termination judgment.

A convergence criterion is set for the iterative process. After calculating the objective's value for the present selected combinations, it is compared with the previous value. If the relative error is larger than 0 but less than ε, terminate the iterative process and output the optimal design. Otherwise, go to the next step of space reduction and perform the next iterative process.

The space reduction method is defined as follows for the transition to the next iteration step. Define $[a, b]$ as the initial space of one control factor and L as the step size. Besides, the step size of each control factor will be halved in the next iterative process. Assume the best value of the control factor is x_0 and each factor has three levels, then the design levels of the next iterative process will be calculated by

$$\begin{cases} (a, a+L/2, a+L), & x_0 - L/2 < a \\ (b-L, b-L/2, b), & x_0 + L/2 > b\,. \\ (x_0 - L/2, x_0, x_0 + L/2), & \text{others} \end{cases} \tag{7.1}$$

In the same way, if each factor has four levels, the space reduction formula is:

$$\begin{cases} (a, a+L/2, a+L, a+3L/2), & x_0 - 3L/4 < a \\ (b-3L/2, b-L, b-L/2, b), & x_0 + 3L/4 > b\,. \\ (x_0 - 3L/4, x_0 - L/4, x_0 + L/4, x_0 + 3L/4), & \text{others} \end{cases} \tag{7.2}$$

7.3 Comparison Between Component-Level and System-Level Robust Optimizations

7.3.1 *Robust Optimization Process*

This example investigates a segmented-rotor SRM (SRSRM) presented in the last chapter. The FEM is presented in Fig. 7.2.

Three objectives, i.e., torque, loss, and torque ripple, are selected as the optimization objectives since they are the main obstacles to the wide application of SRMs in the industry. These three objectives can be combined into one optimization function defined as

$$\begin{aligned} \min: & \; F(\mathbf{x}_s) = w_1 \frac{T_{\text{avg\_initial}}}{T_{\text{avg}}} + w_2 \frac{P_{loss}}{P_{\text{loss\_initial}}} + w_3 \frac{T_{\text{ripple}}}{T_{\text{ripple\_initial}}} \\ \text{s.t.} & \; g_1(\mathbf{x}_s) = sf - 0.6 \le 0 \\ & \; g_2(\mathbf{x}_s) = T_{\text{coil}} - 75 \le 0 \\ & \; \mathbf{x}_{sl} \le \mathbf{x}_s \le \mathbf{x}_{su} \end{aligned} \tag{7.3}$$

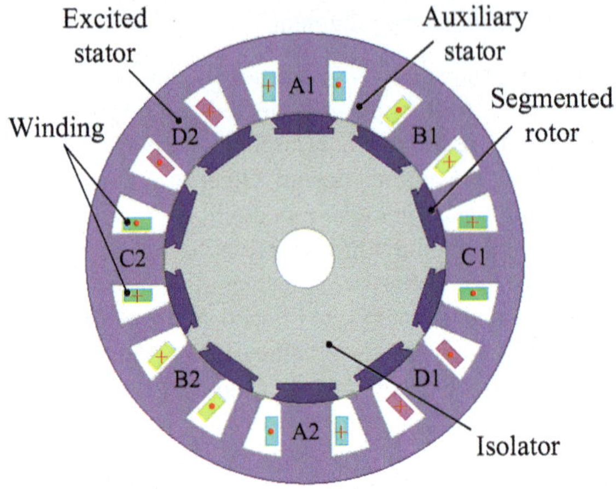

Fig. 7.2 The finite element model of the SSRM

where $T_{\text{avg\_initial}}$, $P_{\text{loss\_initial}}$, and $T_{\text{ripple\_initial}}$ are the average torque, loss, and torque ripple of the initial design, respectively; $\mathbf{x}_s$ is the vector space of the design variables, and g_i the constraint; $\mathbf{x}_{sl}$ and $\mathbf{x}_{su}$ are the lower and upper boundaries, respectively; sf and T_{coil} (in °C) represent the slot fill factor and the temperature of the coil, respectively, and w_1, w_2, and w_3 are the weighting factors. In this example, the torque is set as the prime objective, followed by the loss and torque ripple. Thus, w_1, w_2, and w_3 are assigned 0.4, 0.3, and 0.3, respectively.

Figure 7.3 presents a thermal network model for the investigated SSRM. R_h, R_{sy}, R_{st}, R_r, and R_g are the equivalent thermal resistances of the housing, the stator yoke, the stator tooth, the rotor, and the air gap, respectively. The heat sources in this model include the stator yoke core loss (P_{sy}), the stator teeth core loss (P_{st}), the rotor core loss (P_r), and the copper loss (P_{cu}), which can be obtained from FEM during the optimization process.

The sensitivity of each design variable on the three optimization objectives can be calculated based on global sensitivity analysis.

Figure 7.4 shows sensitivity analysis results for the SRM drive system. The angular position control (APC) method is utilized in this example. In terms of the comprehensive influence on the three optimization objectives, seven significant parameters, i.e., θ_{on}, θ_{off}, g, n, β_r, β_{s1}, and D_{ro}, are selected as the control factors.

Table 7.1 lists the ranges for control factors. To find the noise factors of this structure, a criterion is proposed to determine the influence of uncertainties:

$$N(x_i) = S(x_i) \cdot \frac{\Delta x_i}{x_{i\_\text{initial}}}. \tag{7.4}$$

where Δx_i is the manufacturing tolerance of x_i, and x_{i_initial} is the initial value of x_i.

Figure 7.5 illustrates the influence of uncertainties obtained by (7.4). As shown, the uncertainty of the air gap g has a high impact on the performance of this type of motor, whereas the others only have little influence because the air gap g exhibits

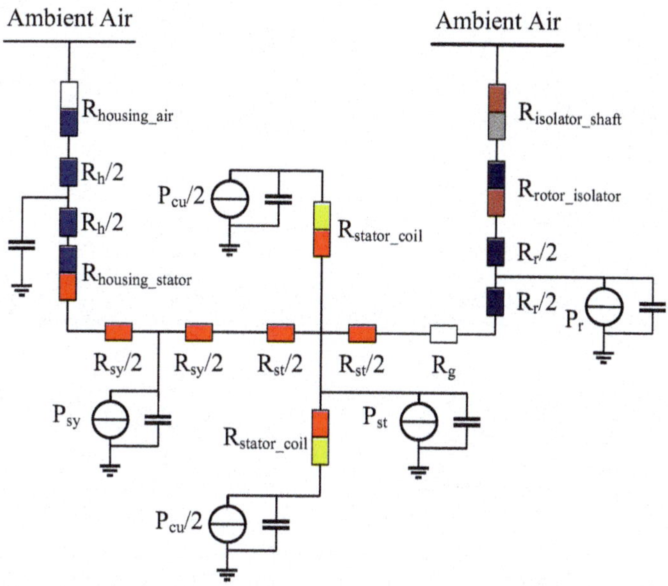

Fig. 7.3 Thermal network model of the SRSRM

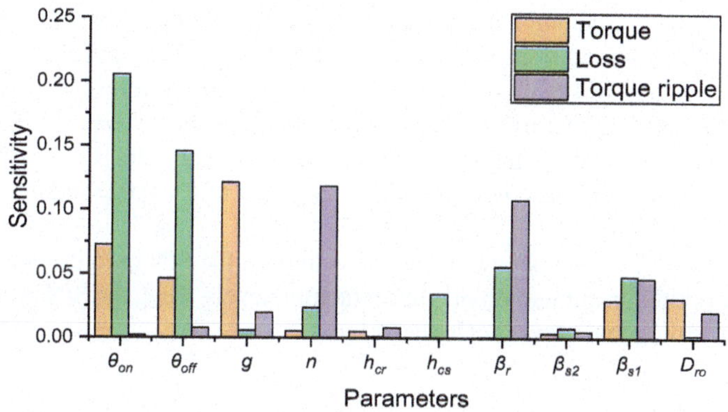

Fig. 7.4 Sensitivity indices

Table 7.1 Control factors and ranges

Control factor	Unit	Ranges
θ_{on}	°	[−4, −2]
θ_{off}	°	[10, 12]
n	–	[20, 26]
D_{ro}	mm	[75, 85]
β_{s1}	°	[18, 21.5]
β_r	°	[24, 30]
g	mm	[0.25, 0.5]

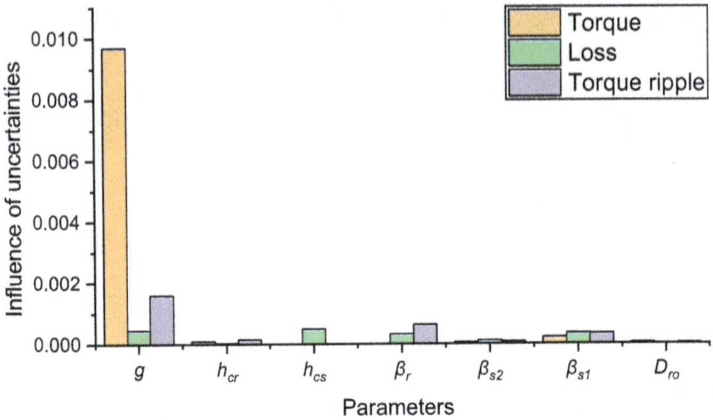

Fig. 7.5 Influence of uncertainties

the highest value of $\Delta x_i/x_{i\_initial}$. Thus, g is determined as the noise factor with the two levels of −0.02 and +0.02 mm, where −0.02 mm means subtracting 0.02 mm from the initial value and + 0.02 mm means adding 0.02 mm to the initial value.

An orthogonal array generated from the control factors is listed in Table 7.2. The numbers in this table correspond to the control factor levels listed in Table 7.3. For one iteration, 36 (18 × 2) combinations of finite element simulations will be performed where 18 is the row number of the orthogonal array for the control factors, and 2 is the combination of the noise factor corresponding to the levels of −0.02 and + 0.02 mm. Taking the first row in Table 7.2 as an example, the values of θ_{on}, θ_{off}, n, D_{ro}, β_{s1}, and β_{r} corresponding to their levels are −4°, 11°, 26, 75 mm, 21.5°, and 30°, respectively, and g has two values, i.e., 0.48 mm (0.5 −0.02) and 0.52 mm (0.5 +0.02).

According to the data obtained from the samples, the S/N ratio can be computed to find the optimal combination of control factors' levels. First, the S/N ratio for each row of the orthogonal array is calculated by

$$SN(i) = -10 \times \lg\left[\frac{1}{2} \times \sum_{j=1}^{2} F^2(i,j)\right] \tag{7.5}$$

where lg means the logarithm operator and i and j are the row numbers in Table 7.2 and the index of the noise factor, respectively. Then, the average S/N ratios for all levels of optimization variables based on the S/N ratios of each row are computed. For example, the average S/N ratio for the first level of θ_{on} is the average value of $SN(1)$, $SN(2)$, $SN(4)$, SN(7), $SN(10)$ and $SN(14)$. The S/N ratios can determine the combination of the best design levels for the optimization variables. The higher S/N ratio means higher robustness. Thus, the level of each optimization variable, which reveals the highest S/N ratio, is selected.

Table 7.2 Orthogonal array of SLSTM

No.	θ_{on}	θ_{off}	n	D_{ro}	β_{s1}	β_r	g
1	1	2	3	1	3	3	3
2	1	3	2	3	1	3	2
3	3	3	3	3	3	2	1
4	1	3	3	2	2	1	3
5	3	3	1	1	2	3	1
6	2	2	2	2	2	3	1
7	1	2	1	3	2	2	2
8	3	1	2	3	2	1	3
9	2	3	2	1	3	1	2
10	1	1	1	1	1	1	1
11	3	2	2	1	1	2	3
12	3	2	1	2	3	1	2
13	2	1	1	3	3	3	3
14	1	1	2	2	3	2	1
15	2	3	1	2	1	2	3
16	3	1	3	2	1	3	2
17	2	1	3	1	2	2	2
18	2	2	3	3	1	1	1

By using SLSTM, the S/N ratios of each iteration are illustrated in Fig. 7.6. As shown in Fig. 7.6a, the best levels in iteration 1 for the seven control factors (θ_{on}, θ_{off}, n, D_{ro}, β_{s1}, β_r, and g) are 2, 1, 2, 3, 3, 3, and 1, which are corresponding to −3°, 10°, 23 mm, 85 mm, 21.5°, 30° and 0.25 mm, respectively. Taking θ_{on} as an example to explain (7.1) in more detail, the best value in iteration 1 is −3°, as illustrated above. As the initial step is 1°, the next step is 0.5°. Since −3.5° (−3 −0.5) > −4° and −2.5° (−3 + 0.5) < −2°, the next three levels of this control factor will be −3.5°, −3°, and −2.5°, respectively.

The best values of each parameter are presented in bold in Table 7.3. The values of the optimization function F of the three iterations are 1.0502, 0.8670, and 0.8620, respectively. The relative error between Iterations 2 and 3 is 0.58% (less than ε). Thus, the iterative process will be terminated after the third iteration.

Table 7.3 Levels for control factors of SLSTM in each iteration

Control factor	Level	Iteration 1	Iteration 2	Iteration 3
θ_{on} (°)	1	−4	−3.5	−3.3
	2	**−3**	**−3**	**−3**
	3	−2	−2.5	−2.8
θ_{off} (°)	1	**10**	**10**	**10**
	2	11	10.5	10.3
	3	12	11	10.5
n	1	20	**21**	**21**
	2	**23**	23	22
	3	26	25	23
D_{ro} (mm)	1	75	80	82.5
	2	80	82.5	83.8
	3	**85**	**85**	**85**
β_{s1} (°)	1	18	19.8	20.6
	2	19.8	20.6	21.1
	3	**21.5**	**21.5**	**21.5**
β_r (°)	1	24	**27**	**27**
	2	27	28.5	27.8
	3	**30**	30	28.5
g (mm)	1	**0.25**	0.25	**0.31**
	2	0.38	0.31	0.35
	3	0.5	**0.38**	0.38

The best values of each parameter are presented in bold.

7.3.2 *Comparison Between Component-Level and System-Level Optimization*

To better show the effectiveness of the proposed SLSTM, the component-level sequential Taguchi method (CLSTM) is performed for comparison. The only difference between the SLSTM and CLSTM is that the control factors selected from the control level are not included in CLSTM. Thus, in this example, θ_{on} and θ_{off} are excluded in CLSTM.

The optimization flowchart of CLSTM is the same as that of the SLSTM. The levels for control factors of CLSTM are tabulated in Table 7.4. During one iteration, 32 (16 × 2) combinations of samples will be carried out where 16 is the row number of the orthogonal array for the control factors, and 2 is the combination with the noise factor corresponding to the levels of −0.02 and + 0.02 mm. Corresponding selected results of each control factor for each iteration are presented in bold in Table 7.4. The values of the optimization function F of the three iterations are 1.1915, 1.0325,

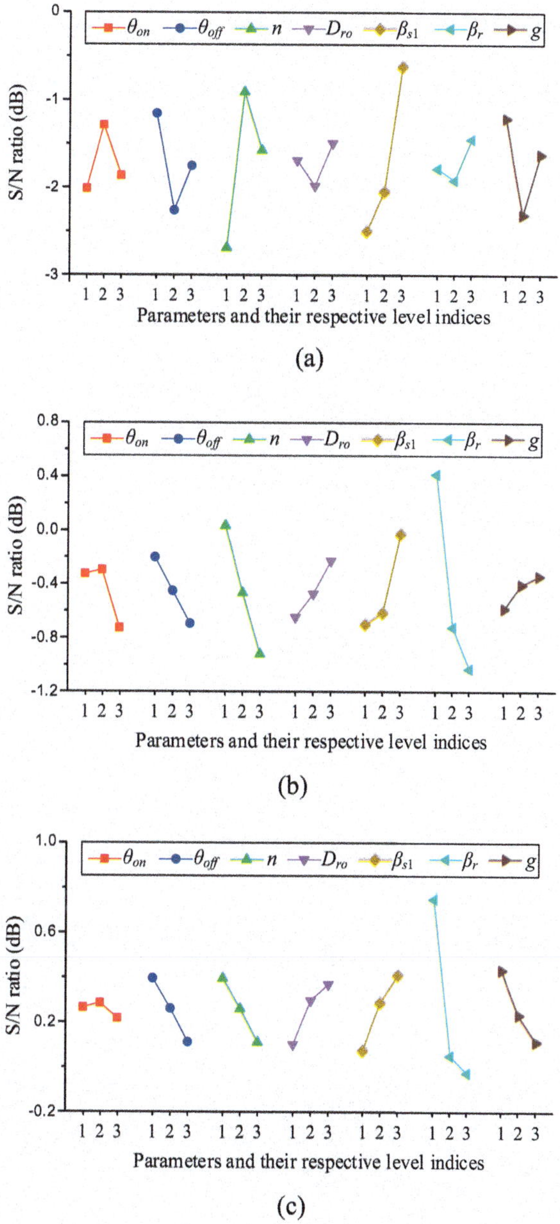

Fig. 7.6 S/N ratios of different iterations for the SSRM by using SLSTM, **a** iteration 1, **b** iteration 2, and **c** iteration 3

and 1.0228, respectively. The relative error between Iterations 2 and 3 is 0.94% (less than ε). Thus, the iterative process will be terminated after the third iteration.

The comparison of torque curves among the initial, CLSTM and SLSTM designs is presented in Fig. 7.7. Detailed comparisons are presented in Table 7.5. Moreover, to further compare the robustness performance of the three designs, a set of input data about the air gap that conforms to the normal distribution after verification is randomly generated. The data distribution in each interval is presented in Fig. 7.8. The FEMs are established according to the different values of the air gap. After the simulation, the corresponding output results of F are presented in Fig. 7.9.

The following conclusions can be drawn.

(1) For the CLSTM, the average torque, loss, torque ripple, and objective value (represented by the value of optimization function F) are 3.30 N m, 225.45 W, 96.28%, and 1.0228, respectively. Compared with the initial design, the average torque is increased by 8.91%, the loss is increased by 15.58%, and the torque ripple also increases slightly. The objective value is reduced by 2.28%. It can be explained that since the sequential Taguchi method aims to find a robust solution

Table 7.4 Levels for control factors of CLSTM in each iteration

Control factor	Level	Iteration 1	Iteration 2	Iteration 3
n	1	20	**22**	**22**
	2	22	23	23
	3	**24**	25	23
	4	26	26	24
D_{ro} (mm)	1	75	75.8	78.3
	2	**78.3**	77.5	79.1
	3	81.7	79.1	80
	4	85	**80.8**	**80.8**
β_{s1} (°)	1	18	19.8	20.6
	2	19.2	20.3	20.9
	3	20.3	20.9	21.2
	4	**21.5**	**21.5**	**21.5**
β_r (°)	1	24	**27**	27
	2	26	28	**27.5**
	3	28	29	28
	4	**30**	30	28.5
g (mm)	1	0.25	0.27	0.33
	2	**0.33**	0.31	**0.35**
	3	0.42	0.35	0.37
	4	0.5	**0.39**	0.39

Corresponding selected results of each control factor for each iteration are presented in bold.

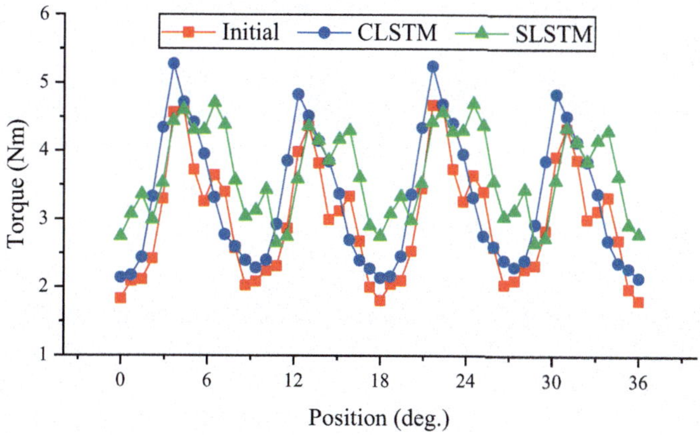

Fig. 7.7 Torque comparison for different designs

Table 7.5 Comparison between initial, CLSTM, and SLSTM designs

Parameter	Unit	Initial	CLSTM	SLSTM
θ_{on}	°	−3	−3	−3
θ_{off}	°	12	12	10
n	–	24	22	21
D_{ro}	mm	82.00	80.8	85
β_{s1}	°	21.38	21.5	21.5
β_r	°	26.64	27.5	27
g	mm	0.25	0.35	0.31
T_{avg}	N m	3.03	3.30	3.66
P_{loss}	W	195.06	225.45	218.17
T_{ripple}	%	93.64	96.28	61.08
F	–	1.00	1.0228	0.8620

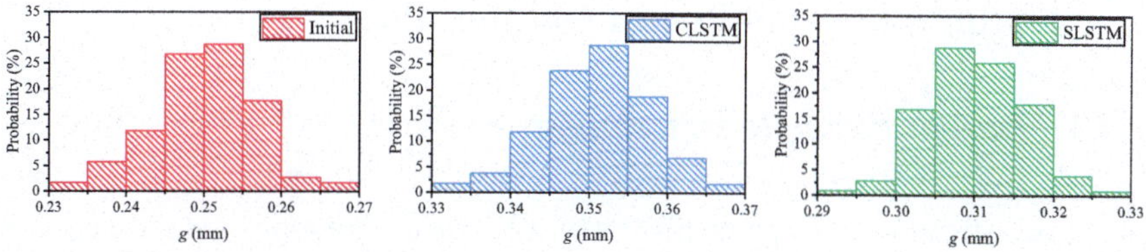

Fig. 7.8 Input data distribution of air gap for different designs

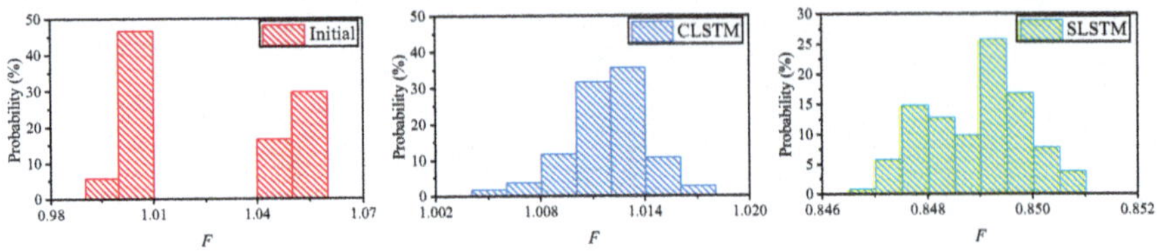

Fig. 7.9 Output results of F for different designs

by selecting levels with high S/N ratios, some performances will be sacrificed to some extent. Thus, it is reasonable that the value of *F* is higher than that of the initial design.

(2) For the SLSTM, the average torque, loss, torque ripple, and objective values are 3.66 N m, 218.17 W, 61.08%, and 0.8620, respectively. Compared with the initial design, the average torque is increased by 20.79% with the sacrifice of 11.80% loss, and the torque ripple is greatly reduced by 34.77%. Therefore, the overall performance has been improved by 13.8%. Besides, the SLSTM exhibits higher average torque, less loss, and lower torque ripple than CLSTM. The SLSTM exhibits the best overall output performance among the three designs.

(3) Considering the robustness performance, from Fig. 7.9, it can be concluded that the initial design exhibits the lowest robustness since *F* is distributed at both ends of the maximum and minimum values and achieves a range of 0.08, while those of CLSTM and SLSTM are only 0.013 and 0.005, respectively. The distributions of *F* in CLSTM and SLSTM are more concentrated and conform to the normal distribution, which can verify the effectiveness of the sequential Taguchi method. Moreover, compared with the CLSTM, the distribution of *F* in SLSTM is narrower. This implies that the proposed SLSTM can achieve a higher level of robustness for SRM drive systems.

7.4 Comparison Between System-Level Deterministic Optimization and Robust Optimization

7.4.1 Design Example Description

In this investigation, a six-phase SRM with 12/10 stator/rotor poles is selected as an example, as shown in Fig. 7.10. The detailed parameter diagram is shown in Fig. 7.11. The design specifications of the motor are tabulated in Table 7.6. Table 7.7 lists the main design parameters and their ranges. The structural parameters of the motor are selected from *D*, α_s, α_r, L_{sy}, L_{ry}, and g, where α_s and α_r are the stator and rotor pole arc coefficients, respectively.

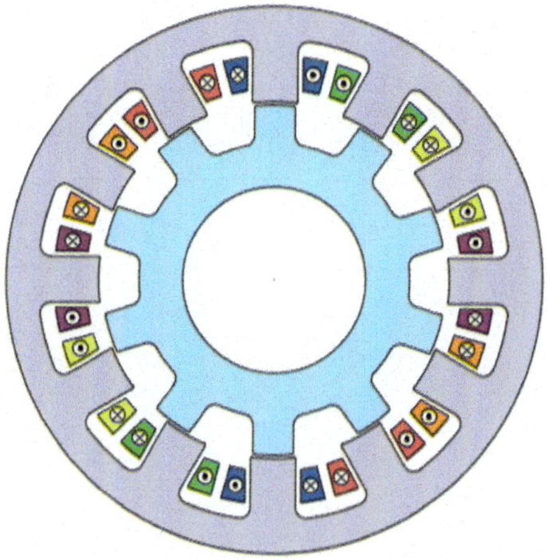

Fig. 7.10 Topology of the six-phase 12/10 SRM

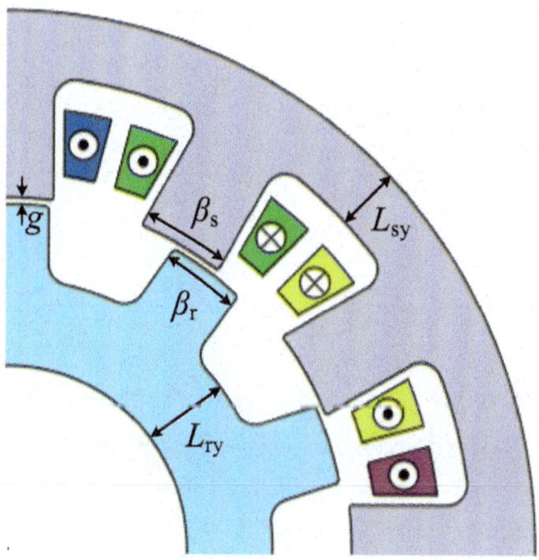

Fig. 7.11 Schematic diagram of motor parameters

In this example, the output torque, loss (the sum of the copper loss and iron loss), and torque ripple are selected as the optimization objectives, aiming to maximize the output torque while minimizing loss and torque ripple. The weight coefficient is introduced to represent the proportion of each target and independent constraints are set simultaneously. An optimization model including these three optimization objectives is established as

Table 7.6 Design specifications

Parameters	Unit	Value
Rated speed n	r/min	3000
Rated voltage U	V	24
Rated torque T_{avg}	N m	0.8
Torque ripple T_{rip}	%	< 30
Efficiency η	%	> 82.5
Number of coils N	–	25
Stator outer diameter D_{so}	mm	80
Rotor inner diameter D_{ri}	mm	29
Axial length l	mm	70

Table 7.7 Initial values and ranges of optimization variables

Design parameters	Unit	Initial value	Range
Rotor outer diameter D	mm	46	44–50
Stator pole arc coefficient α_{s}	–	0.45	0.4–0.5
Rotor pole arc coefficient α_{r}	–	0.35	0.3–0.4
Stator yoke width L_{sy}	mm	4.6	4.4–5
Rotor yoke width L_{ry}	mm	3.8	3.6–4.2
Air gap g	mm	0.3	0.3–0.5
Turn-on angle θ_{on}	°	−2	−4–−1
Conduction angle θ_{c}	°	14	13–16

$$\begin{aligned} &\min : F = \lambda_1 \frac{T_{\text{avg\_0}}}{T_{\text{avg}}} + \lambda_2 \frac{P_{\text{loss}}}{P_{\text{loss\_0}}} + \lambda_3 \frac{T_{\text{rip}}}{T_{\text{rip\_0}}} \\ &\text{s.t.} \quad \begin{cases} g_1(\mathbf{x}_s) = 0.6 - T_{\text{avg}} \leq 0 \\ g_2(\mathbf{x}_s) = T_{\text{rip}} - 30 \leq 0 \\ g_3(\mathbf{x}_s) = 82.5 - \eta \leq 0 \\ g_4(\mathbf{x}_s) = sf - 0.6 \leq 0 \\ \mathbf{x}_{sl} \leq \mathbf{x}_s \leq \mathbf{x}_{su} \end{cases} \end{aligned} \tag{7.6}$$

where T_{avg}, P_{loss}, and T_{rip} are the average torque, loss, and torque ripple, respectively; $\text{T}_{\text{avg\_0}}$, $\text{P}_{\text{loss\_0}}$, and $\text{T}_{\text{rip\_0}}$ are the initial values of these three optimization objectives, respectively; λ_1, λ_2, and λ_3 are the weight coefficients, and sf is the slot fill factor; $\mathbf{x}_{\text{s}}$ is the design parameter vector, and $\mathbf{x}_{\text{sl}}$ and $\mathbf{x}_{\text{su}}$ are the lower and upper limits of $\mathbf{x}_{\text{s}}$, respectively. In this example, to balance the performance of the optimization objective, $\lambda_1 = \lambda_2 = \lambda_3 = 1/3$ is taken.

Table 7.8 Sensitivity analysis results

Parameter	Sensitivity			
	Torque	Loss	Ripple	Final
D	0.05	0.02	0.05	0.04
α_s	0.17	0.05	0.04	0.09
α_r	0.12	0.13	0.04	0.10
L_{sy}	0.02	0.01	0.05	0.03
L_{ry}	0.01	0.01	0.05	0.02
g	0.03	0.16	0.46	0.22
θ_{on}	0.29	0.25	0.04	0.19
θ_c	0.30	0.35	0.06	0.24

To obtain the comprehensive sensitivity of the parameters, a weighted analysis is carried out on the sensitivity values of the three optimization objectives. The weighting coefficient is consistent with (7.6), and the sensitivity results are obtained by global sensitivity analysis. Table 7.8 lists the sensitivity analysis results. According to the comprehensive sensitivity, all variables whose comprehensive sensitivities are not less than 0.1 are selected as optimization variables. Therefore, g, θ_{on}, θ_c, α_s, and α_r are the optimization variables for deterministic and robust optimization, and the rest of the variables are kept fixed at their initial values.

7.4.2 Deterministic Optimization

The Pareto solutions of the deterministic optimization are presented in Fig. 7.12. The parameters have been divided into two levels, X_1 and X_2. According to the sensitivity results, X_1 includes g, θ_{on}, and θ_c, and X_2 includes α_s and α_r. Among the optimal solutions in each level, the deterministic solution is selected according to (7.6), and the results are presented in Table 7.9. After two iteration processes, the values of F converge.

7.4.3 Robust Optimization

The sequential Taguchi method is applied to the robust optimization. The airgap length g is selected as the noise factor among the five optimization variables according to the method proposed in the last section. Four levels are chosen for each optimization parameter, and thus, the space of the next iteration can be calculated by (7.2).

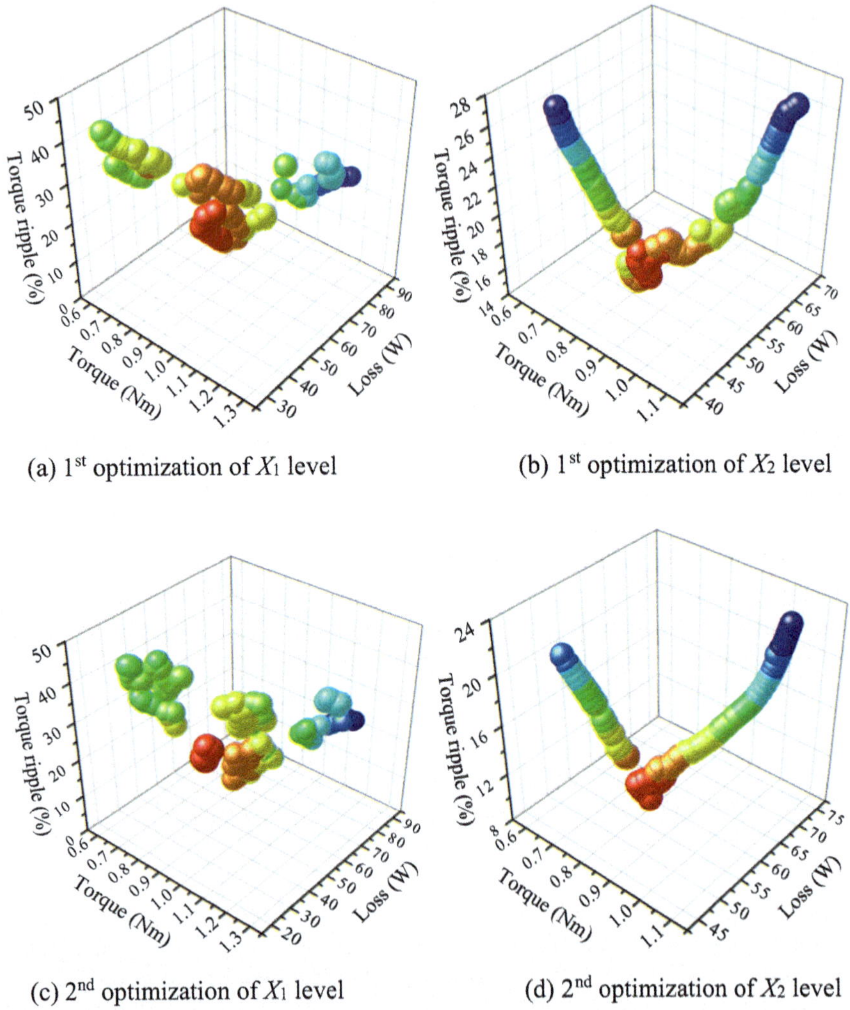

Fig. 7.12 Pareto solutions of the deterministic optimization

The S/N ratio of each iteration is presented in Fig. 7.13. A higher S/N ratio corresponds to better robustness, so the optimized parameters in each iteration are the parameter values corresponding to a high S/N ratio. Table 7.10 lists the robust optimization results of each iteration. It can be found that the convergence condition is satisfied after 4 iterations, and the results of the 3rd and 4th iterations are the same.

Table 7.9 Iteration results of the deterministic approach

Parameter	Iteration	
	1st	2nd
g	0.36	0.42
θ_{on}	−1.2	−1.1
θ_c	16	15.9
α_s	0.4	0.4
α_r	0.4	0.4
T_{avg} (N m)	0.85	0.84
P_{loss} (W)	50.84	56.04
T_{rip} (%)	15.83	10.34
F	0.88	0.88

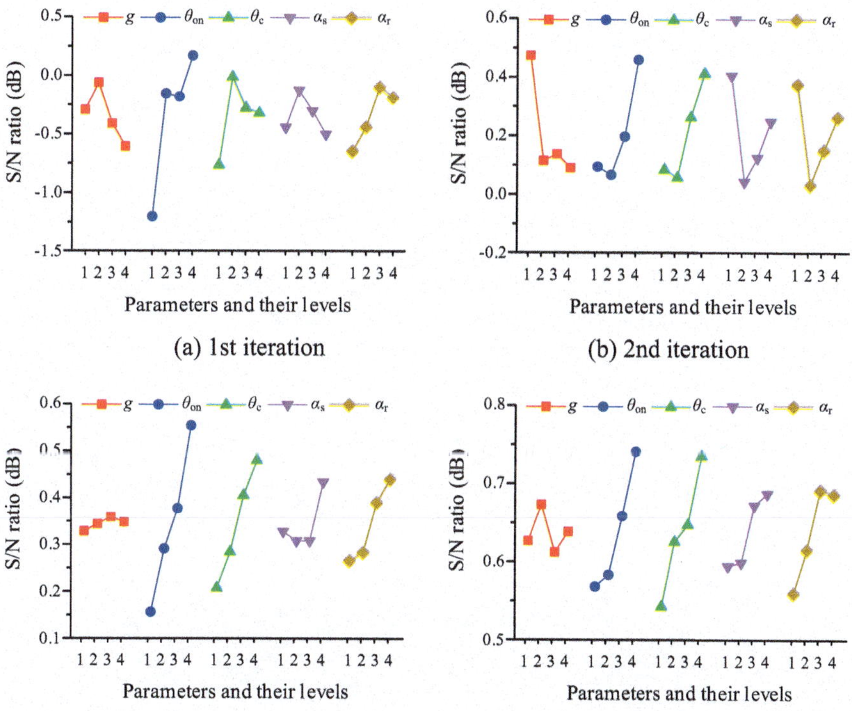

(a) 1st iteration

(b) 2nd iteration

(c) 3rd iteration

(d) 4th iteration

Fig. 7.13 Signal/Noise values in robust approach

Table 7.10 Iteration results of the robust approach

Parameter	Iteration		
	1st	2nd	3rd/4th
g	0.37	0.32	0.35
θ_{on}	−1	−1	−1
θ_c	14	14.8	14.8
α_s	0.43	0.41	0.44
α_r	0.37	0.35	0.38
T_{avg} (N m)	0.75	0.89	0.77
P_{loss} (W)	39.66	44.50	41.19
T_{rip} (%)	30.73	31.47	23.26
F	0.96	0.94	0.90

7.4.4 Comparison Between Deterministic and Robust Optimization

Figure 7.14 presents the torque comparison among the initial, deterministic, and robust designs after optimization. The initial design suffers from a large torque ripple, which cannot meet the design requirements. The optimization results are tabulated in Table 7.11. It can be found that deterministic optimization greatly reduces the torque ripple and obtains the lowest F value, which exhibits the best performance. In contrast, the torque characteristics of the optimal solution obtained by robust optimization are not as good as those obtained by deterministic optimization.

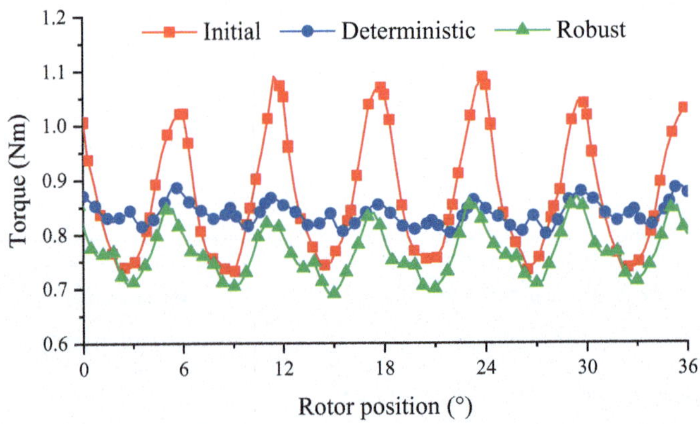

Fig. 7.14 Output torque comparison

Table 7.11 Initial design and optimization results

Parameter	Initial	Deterministic	Robust
g	0.3	0.42	0.35
θ_{on}	−2	−1.1	−1
θ_c	14	15.9	14.8
α_s	0.45	0.4	0.44
α_r	0.35	0.4	0.38
η (%)	87	82.5	85.4
T_{avg} (N m)	0.87	0.84	0.77
P_{loss} (W)	40.82	56.04	41.19
T_{rip} (%)	41.54	10.34	23.26
F	1	0.88	0.90

Assume that the noise factor follows a normal distribution with standard deviation σ_i, where $\sigma_i = \Delta x_i/3$. After Monte Carlo analysis and corresponding simulation model setting, the performance distribution diagram of the design point with respect to the noise factor can be obtained, which considers the performance distribution characteristics of the mass production of motors under the influence of errors. Figures 7.15, 7.16 and 7.17 show the performance distribution diagrams of the initial motor design, deterministic design, and robust design, respectively. In this example, the product failure rate P_f is introduced to express the product rejection rate and measure the reliability of the product. Assuming that all constraints are independent of each other, then P_f is defined as

$$P_f = 1 - \prod_{i=1}^{m=4} P\{g_i(\mathbf{x}_s) \leq 0\}. \tag{7.7}$$

From Fig. 7.15, it can be found that the P_f value of the initial design is 100% due to the high torque ripple. From the efficiency distribution diagram in Fig. 7.16, the efficiencies of some deterministic designs are lower than 82.5%, and thus its P_f value reaches 64.8%. Besides, the loss of the deterministic designs is relatively high, resulting in a large current density, which is not conducive to the heat dissipation of the motor. Referring to Fig. 7.17, the performance distribution of the robust optimization is within the allowable range. Thus, its P_f value is 0, ensuring all products meet the requirements.

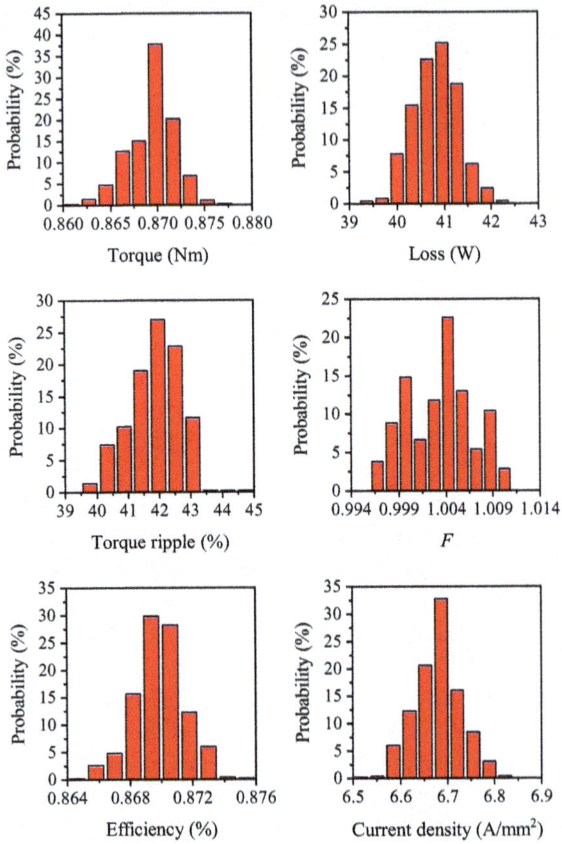

Fig. 7.15 Performance distribution of initial design

Moreover, comparing the distribution of F value in these figures, the ranges of F value distribution in the initial design, deterministic design, and robust design are 0.014, 0.023, and 0.005, respectively. It can be found that the distribution of F value in robust optimization is more concentrated. Therefore, the overall performance of the optimal solution is the most robust.

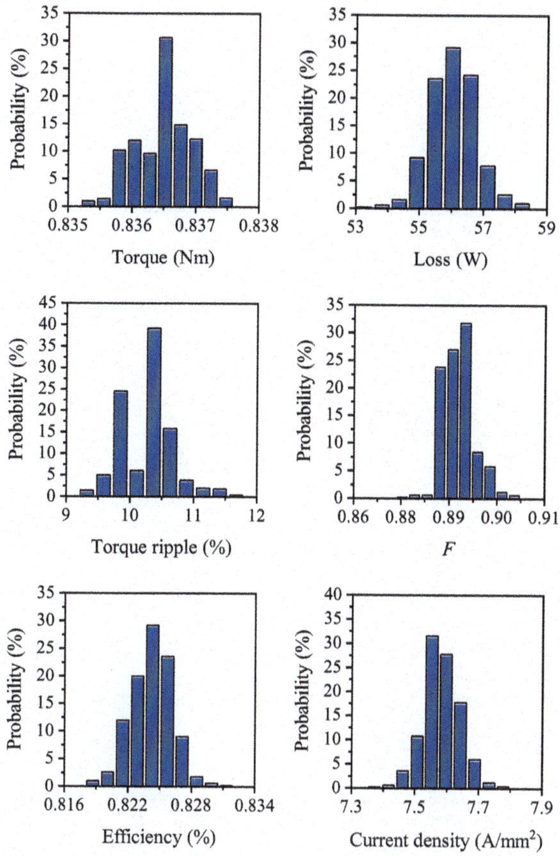

Fig. 7.16 Performance distribution of deterministic approach

7.5 Summary

In this chapter, a robust system-level optimization approach based on the sequential Taguchi method is proposed for the SRM drive systems with the consideration of manufacturing tolerances. DoE technique and sensitivity analysis are utilized during the optimization process. The investigation of two case studies shows that SLSTM can significantly improve output performance and system reliability, which will benefit the manufacturing of SRMs and drive systems from the engineering design perspective.

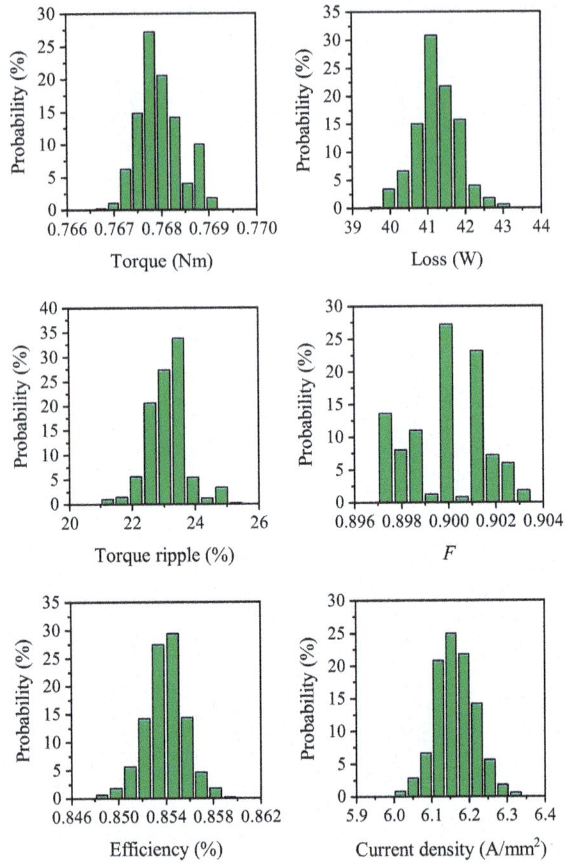

Fig. 7.17 Performance distribution of robust approach

References

1. Bramerdorfer G, Zăvoianu A. Surrogate-based multi-objective optimization of electrical machine designs facilitating tolerance analysis. IEEE Trans Magn. 2017;53(8):1–11. https://doi.org/10.1109/TMAG.2017.2694802.
2. Diao K, Sun X, Lei G, Guo Y, Zhu J. Multimode optimization of switched reluctance machines in hybrid electric vehicles. IEEE Trans Energy Convers. 2020. https://doi.org/10.1109/TEC.2020.3046721.
3. Song S, Zhang M, Ge L, Wang L. Multiobjective optimal design of switched reluctance linear launcher. IEEE Trans Plasma Sci. 2015;43(5):1339–45. https://doi.org/10.1109/tps.2015.2416204.
4. Diao K, Sun X, Lei G, Guo Y, Zhu J. Multiobjective system level optimization method for switched reluctance motor drive systems using finite-element model. IEEE Trans Ind Electron. 2020;67(12):10055–64. https://doi.org/10.1109/tie.2019.2962483.
5. Chen H, Yan W, Gu JJ, Sun M. Multiobjective optimization design of a switched reluctance motor for low-speed electric vehicles with a Taguchi–CSO algorithm. IEEE/ASME Trans Mechatron. 2018;23(4):1762–74. https://doi.org/10.1109/tmech.2018.2839619.

6. Diao K, Sun X, Lei G, Bramerdorfer G, Guo Y, Zhu J. System-level robust design optimization of a switched reluctance motor drive system considering multiple driving cycles. IEEE Trans Energy Convers. 2021;36(1):348–57. https://doi.org/10.1109/TEC.2020.3009408.
7. Lei G, Wang T, Guo Y, Zhu J, Wang S. System-level design optimization methods for electrical drive systems: deterministic approach. IEEE Trans Ind Electron. 2014;61(12):6591–602. https://doi.org/10.1109/TIE.2014.2321338.
8. Sun X, Shi Z, Cai Y, Lei G, Guo Y, Zhu J. Driving-cycle-oriented design optimization of a permanent magnet hub motor drive system for a four-wheel-drive electric vehicle. IEEE Trans Transp Electrification. 2020;6(3):1115–25. https://doi.org/10.1109/TTE.2020.3009396.
9. Li S, Zhang S, Habetler TG, Harley RG. Modeling, design optimization, and applications of switched reluctance machines—a review. IEEE Trans Ind Appl. 2019;55(3):2660–81. https://doi.org/10.1109/TIA.2019.2897965.
10. Davarpanah G, Faiz J. Nonlinear modeling of a C-core connected two-phase switched reluctance motor. IEEE Trans Energy Convers. 2021;36(4):2761–9. https://doi.org/10.1109/tec.2021.3062501.
11. Diao K, Sun X, Lei G, Bramerdorfer G, Guo Y, Zhu J. Robust design optimization of switched reluctance motor drive systems based on system-level sequential Taguchi method. IEEE Trans Energy Convers. 2021;36(4):3199–207. https://doi.org/10.1109/tec.2021.3085668.
12. Diao K, Sun X, Yao M. Robust-oriented optimization of switched reluctance motors considering manufacturing fluctuation. IEEE Trans Transp Electrification. 2022;8(2):2853–61. https://doi.org/10.1109/tte.2021.3137385.
13. Degano M, Carraro E, Bianchi N. Selection criteria and robust optimization of a traction PM-assisted synchronous reluctance motor. IEEE Trans Ind Appl. 2015;51(6):4383–91.
14. Hasegawa Y, Nakamura K, Ichinokura O. Optimization of a switched reluctance motor made of Permendur. IEEE Trans Magn. 2010;46(6):1311–4. https://doi.org/10.1109/TMAG.2010.2041903.
15. Bramerdorfer G, Tapia JA, Pyrhönen JJ, Cavagnino A. Modern electrical machine design optimization: techniques, trends, and best practices. IEEE Trans Ind Electron. 2018;65(10):7672–84. https://doi.org/10.1109/TIE.2018.2801805.
16. Lei G, Liu C, Guo Y, Zhu J. Robust multidisciplinary design optimization of PM machines with soft magnetic composite cores for batch production. IEEE Trans Magn. 2016;52(3):1–4. https://doi.org/10.1109/TMAG.2015.2485983.
17. Diao K, Sun X, Bramerdorfer G, Cai Y, Lei G, Chen L. Design optimization of switched reluctance machines for performance and reliability enhancements: a review. Renew Sust Energ Rev. 2022;168. https://doi.org/10.1016/j.rser.2022.112785.
18. Bramerdorfer G, Lei G, Cavagnino A, Zhang Y, Sykulski J, Lowther DA. More robust and reliable optimized energy conversion facilitated through electric machines, power electronics and drives, and their control: state-of-the-art and trends. IEEE Trans Energy Convers. 2020;35(4):1997–2012. https://doi.org/10.1109/tec.2020.3013190.
19. Lei G, Wang T, Zhu J, Guo Y, Wang S. System-level design optimization method for electrical drive systems—robust approach. IEEE Trans Ind Electron. 2015;62(8):4702–13. https://doi.org/10.1109/TIE.2015.2404305.
20. Lei G, Liu C, Li Y, Chen D, Guo Y, Zhu J. Robust design optimization of a high-temperature superconducting linear synchronous motor based on Taguchi method. IEEE Trans Appl Supercond. 2019;29(2):1–6. https://doi.org/10.1109/TASC.2018.2882426.
21. Lei G, Bramerdorfer G, Ma B, Guo Y, Zhu J. Robust design optimization of electrical machines: multi-objective approach. IEEE Trans Energy Convers. 2021;36(1):390–401. https://doi.org/10.1109/TEC.2020.3003050.
22. Lee S, Kim K, Cho S, Jang J, Lee T, Hong J. Optimal design of interior permanent magnet synchronous motor considering the manufacturing tolerances using Taguchi robust design. IET Elect Power Appl. 2014;8(1):23–8. https://doi.org/10.1049/iet-epa.2013.0109.

23. Omekanda AM. Robust torque and torque-per-inertia optimization of a switched reluctance motor using the Taguchi methods. IEEE Trans Ind Appl. 2006;42(2):473–8. https://doi.org/10.1109/TIA.2006.870031.
24. Lei G, Bramerdorfer G, Liu C, Guo Y, Zhu J. Robust design optimization of electrical machines: a comparative study and space reduction strategy. IEEE Trans Energy Convers. 2021;36(1):300–13. https://doi.org/10.1109/TEC.2020.2999482.

Chapter 8
Application-Oriented Optimization Method for Switched Reluctance Motor Drive Systems

Abstract Application-oriented design optimization methods are a set of techniques used to optimize the design of a system or product based on its intended application. These methods are based on the idea that the design of a product should be tailored to the specific requirements of its intended use rather than being optimized for a generic set of specifications. This chapter aims to present an application-oriented design optimization method for switched reluctance motor (SRM) drive systems by considering multiple operation modes. An example of an SRM applied in an electric vehicle is investigated. More optimization concepts, such as multi-mode optimization and correlation analysis regarding the multiple driving cycles, are introduced for the specific application.

Keywords Application-oriented design optimization method · Electric vehicle · Multi-mode optimization · Switched reluctance motors

8.1 Introduction

Design optimization is a process of finding the best possible design solution within the given constraints and objectives. Conventionally, design optimization methods mainly focus on mathematical algorithms and numerical optimization techniques [1–3]. With the advancements in computer technology, design optimization methods have evolved to include application-oriented approaches that incorporate the specific needs of the industry applications. Application-oriented design optimization methods provide a more practical and realistic approach to meet the application's specific requirements [4–6].

Application-oriented design optimization methods use simulation tools, physical testing, and real-world data to develop optimized design solutions that meet the application's specific needs [7–9]. They provide several advantages over conventional optimization methods. These advantages include:

- Tailored to Specific Needs: Application-oriented design optimization methods can be tailored to meet the specific needs of the industry application. To develop

X. Sun et al., *Multi-objective Design Optimization of Switched Reluctance Motor Drive Systems*, https://doi.org/10.1007/978-981-96-0672-6_8

optimized design solutions that meet the application's specific requirements, these methods consider various factors, including cost, performance, manufacturability, and reliability [10].

- Multi-disciplinary Approach: Application-oriented design optimization methods involve a multi-disciplinary approach considering various factors, including mechanical, electrical, and software design. This approach ensures the optimized design solution is practical and applicable to real-world applications [11].
- Fast Time-to-Market: Application-oriented design optimization methods can significantly reduce the time-to-market for new products. By considering various factors, including cost and manufacturability, these methods can develop optimized design solutions that are feasible and practical to produce [12].
- Cost Savings: Application-oriented design optimization methods can significantly reduce the overall cost of the product development process. By considering various factors, including cost and manufacturability, these methods can develop optimal design solutions that are cost-effective and practical to produce [13].

8.2 Application-Oriented Design Optimization Method

Figure 8.1 illustrates a framework for the proposed application-oriented design optimization method. It includes the following main steps.

Step 1: Problem definition

a. Define the objective function: The objective function should be well-defined according to the application's specific requirements. It should be measurable and quantifiable and reflect the desired outcome of the design process.
b. Define the constraints: Constraints should be defined to ensure the design meets the requirements or limitations. These may include physical limitations, performance specifications, and safety requirements.
c. Identify the design variables: The design variables are the parameters that can be adjusted to achieve the objectives. These variables should be identified and defined clearly to enable effective optimization.

Step 2: Design exploration

a. Generate design alternatives: Design alternatives can be generated through manual exploration, simulation, or other methods, depending on the nature of the problem and the available resources.
b. Evaluate design alternatives: Evaluate each alternative design based on the objective and constraints.
c. Refine design alternatives: Refine the design alternatives based on the evaluation results to eliminate inferior designs and identify promising candidates for optimization.

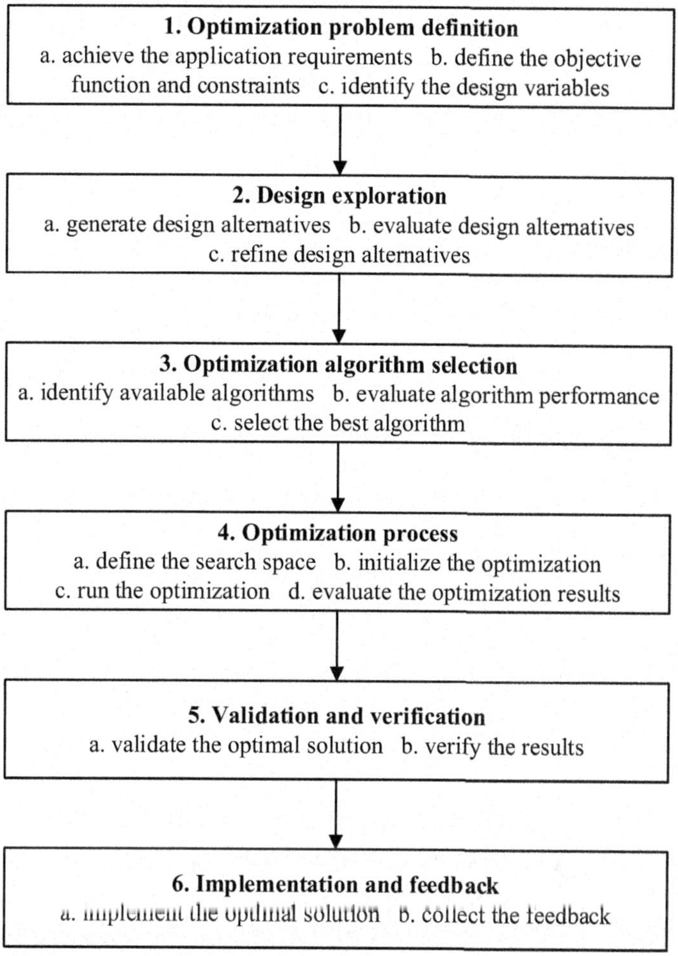

Fig. 8.1 Framework of application-oriented design optimization method

Step 3: Optimization algorithm selection

a. Identify available algorithms: Different optimization algorithms have different strengths. Identify the algorithms that are available and suitable for the problem.
b. Evaluate algorithm performance: Evaluate each algorithm based on its efficiency, effectiveness, and scalability.
c. Select the best algorithm: Select the algorithm best suited for the problem based on the evaluation results.

Step 4: Optimization process

a. Define the search space: Define the search space based on the design variables and constraints.
b. Initialize the optimization: Initialize the optimization with a set of initial designs or parameters.

c. Run the optimization: Run the selected algorithm on the search space to find the optimal solution.
d. Evaluate the optimization results: Evaluate the optimization results based on the objective and constraints. If the results are unsatisfactory, refine the search space and repeat the optimization process.

Step 5: Validation and verification

a. Validate the optimal solution: Validate the optimal solution using physical testing, simulation, or other validation methods to ensure that it meets the desired objective and constraints.
b. Verify the results: Verify the results to ensure they are reliable and accurate.

Step 6: Implementation and feedback

a. Implement the optimal solution: Implement the optimal solution in the real-world application.
b. Collect feedback: Collect feedback from the implementation to identify areas for improvement and refine the optimization process for future applications.

In conclusion, implementing application-oriented design optimization involves a series of well-defined and executed steps, including problem definition, design exploration, optimization algorithm selection, optimization process, validation and verification, and implementation and feedback. Each step is critical in achieving the optimal design that meets the desired objective and constraints.

8.3 System-Level Multi-mode Robust Optimization Method Considering Multiple Driving Cycles

8.3.1 Multiple Driving Cycles and Multi-mode Optimization Method

As reported in [14], the optimization process for the in-wheel SRM in EVs was only conducted under rated operation conditions. However, for the application of EVs, manifold driving conditions are required during operation, such as frequent start and stop, acceleration and deceleration, climbing, and high-speed cruise [15–17]. In [18], the energy efficiency of the traction machine over the driving cycle was characterized by evaluating several representative points. The overall design optimization for SRMs applied in EVs is very challenging in meeting the requirements under different driving conditions [19, 20]. Consequently, SRMs designed for EV application should possess the capability of multi-mode operation.

Figure 8.2 shows the typical new European driving cycles (NEDC) [15]. As shown, several driving modes, i.e., starting and stopping, normal cruise, acceleration, high-speed cruise, and climbing, frequently occur in the NEDC. Thus, it is essential to consider the associated manifold SRM operating points during the optimization process.

Figure 8.3 shows the relationship among the driving modes, design requirements, and motor operation. According to the different driving modes, the designed SRM should meet the requirements for the whole operating range. One driving cycle may put forward multiple design requirements, and one design requirement may be applied to several driving modes. For example, the normal cruise of driving cycles puts forward high torque, low torque ripple, and high motor efficiency. Besides, the low torque ripple is correlated to the driving modes like normal operation, open-circuit fault, and short-circuit fault.

Since different requirements are needed in different driving modes, it is meaningful to propose a multi-objective optimization method to optimize the multiple

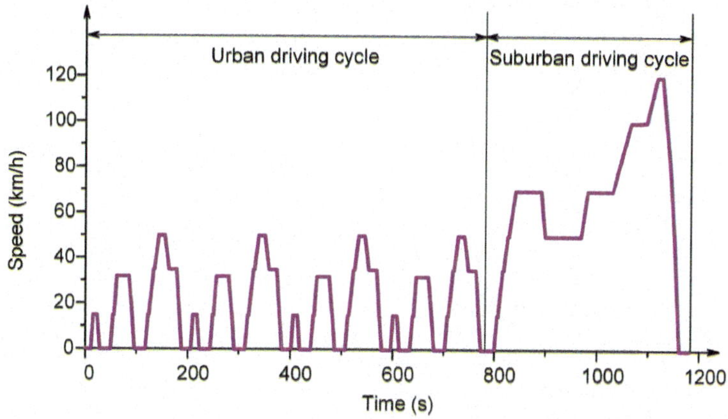

Fig. 8.2 NEDC driving cycle

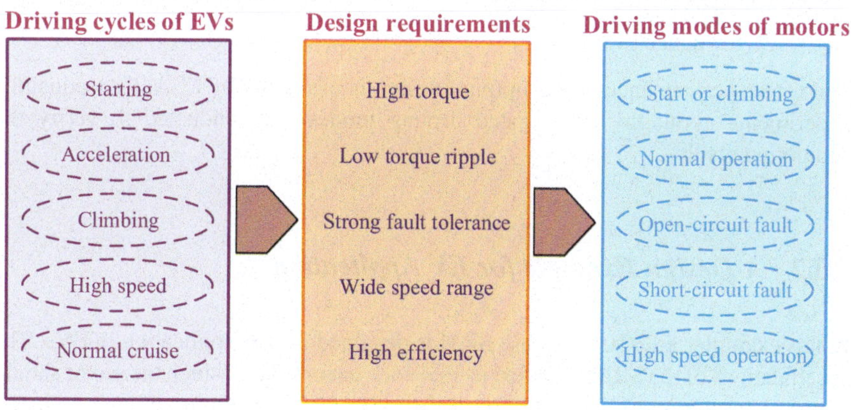

Fig. 8.3 Relationships among driving cycles, design requirements, and the operating range of SRMs

driving modes [21]. Besides, for the application in EVs, the performance evaluation of the motor drive system is necessary, and the manufacturing tolerance should also be considered during the whole optimization process [22]. Thus, this section proposes a design optimization method that integrates the concept of multiple driving modes, system-level optimization, and robust design.

Figure 8.4 shows a procedure for the proposed new system-level robust optimization method for electrical drive systems applied in EVs. This kind of system-level robust optimization method considers the driving cycles of EVs, the geometry and control parameters of the SRM drive system, and manufacturing tolerances. It can be divided into four steps.

Step 1: Initial design

In this step, the driving cycles of EVs are determined first according to road conditions. Then, the requirements of EVs can be defined. Based on the driving cycles and requirements, the operation of the motor can be classified into different modes. Finally, an initial design can be derived based on the requirements of each mode.

Step 2: Sensitivity analysis

Compared with the optimization for one operating point, the proposed approach includes one sensitivity analysis per each driving mode. The significant variables of the drive system in each mode can be selected according to the sensitivity analysis results.

Step 3: Taguchi robust optimization

This step first defines the objective function for all the driving modes. The levels of optimization variables and noise factors are determined to set up the orthogonal array. Then, the required evaluations based on finite element models (FEMs) are specified according to the orthogonal array, which helps significantly reduce the overall number of necessary simulations. Finally, the optimal design can be selected based on the analysis of the results.

Step 4: Performance evaluation

The motor performance of the optimal design is evaluated by FEA. Consequently, experiments about the investigated driving modes are conducted to verify the simulation results.

8.3.2 A Design Example for EV Application

In this example, a 12/10 six-phase SRM is designed for the application in EVs. The specifications of the SRM are listed in Table 8.1, according to which the motor should feature a torque of 120 N m at the rated speed of 6000 rev/min while providing a minimal efficiency of 85%. Figure 8.5 shows the prototype and the assembly process of the SRM. The SRM is mainly comprised of a stator core with 12 poles wrapped

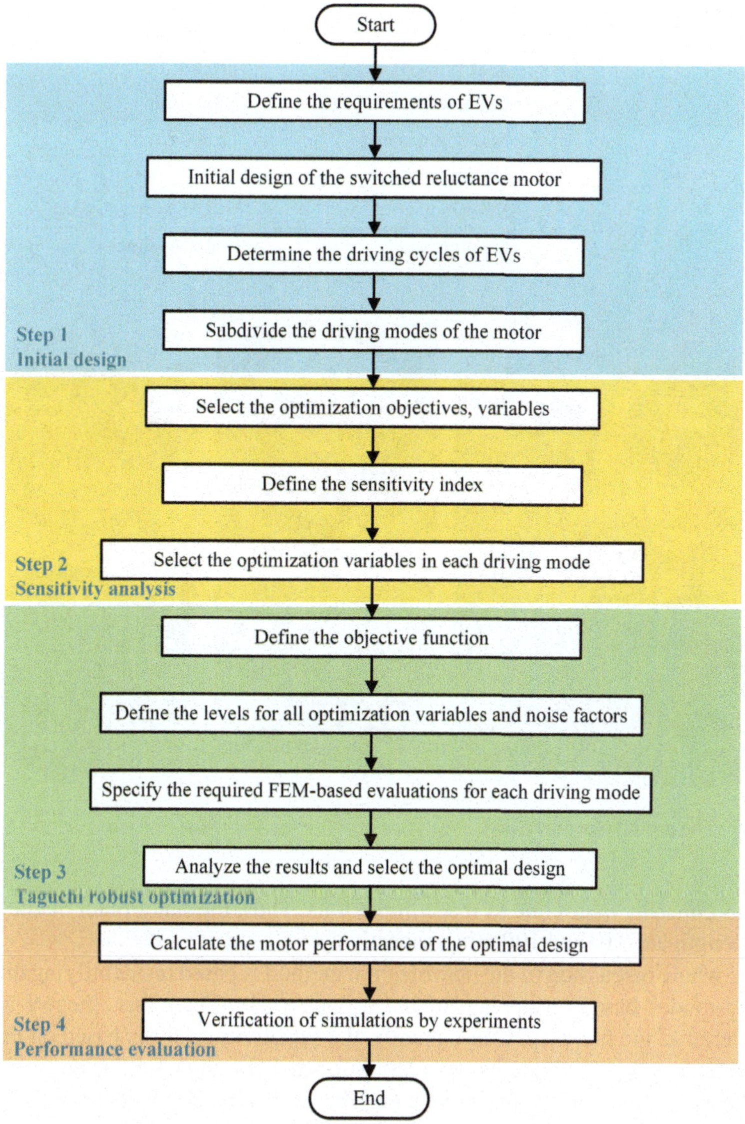

Fig. 8.4 A procedure for the proposed system-level robust optimization method for SRM drive systems

by windings and a rotor core with 10 poles without any coils, magnets, or squirrel cages. It has six phase windings, and each phase winding consists of two coils wound on the opposite poles of the stator. These two coils are connected in parallel. The initial design values are listed in Table 8.2. The external dimensions of the motor are

Table 8.1 Specifications of the SRM

Rated power	75 kW
Peak power	110 kW
Rated speed	6000 r/min
Maximum speed	12,000 r/min
DC link voltage	384 V
Efficiency	≥ 85% at rated speed
Cooling type	Water-cooling

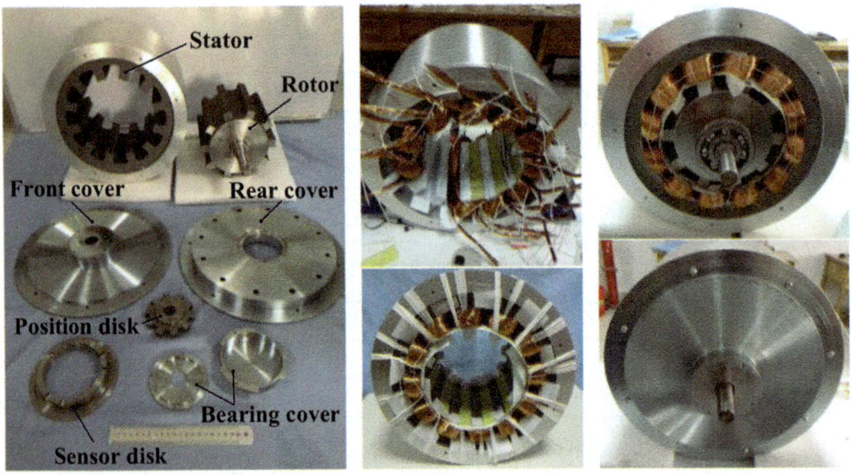

Fig. 8.5 The prototype of the SRM

fixed as constant values due to space limitations. Thus, the stator outer diameter and stack length are 310 and 190 mm, respectively.

The whole discussion of the optimization method is based on identifying different driving modes based on the considered driving cycle. Thus, this categorization is considered as the first step. Consequently, the sensitivity analysis based on FEMs is performed to select the significant variables for each driving mode. Then, since some objectives in different driving modes may be positively correlated, the correlation analysis is conducted to determine these objectives and select only one objective to substitute all the positively correlated objectives, which will significantly reduce the computational cost. Finally, a sequential Taguchi optimization method, which holds the advantage of both the Taguchi method and sequential optimization strategy, is carried out to find the optimal solution for multi-mode applications based on FEMs.

(a) Multi-mode concept

The driving modes of the SRM applied in EVs can be categorized into five types according to the driving cycle and design requirements shown in Figs. 8.2 and 8.3.

Table 8.2 Initial design values

Parameter	Description	Unit	Value
D_{so}	Stator outer diameter	mm	310
l_s	Stack length	mm	190
D_{ro}	Rotor outer diameter	mm	204
β_s	Stator pole arc	°	13.5
β_r	Rotor pole arc	°	12
h_{cs}	Stator yoke thickness	mm	21
h_{cr}	Rotor yoke thickness	mm	20
g	Length of air gap	mm	0.5
n	Number of turns	–	26
θ_{on}	Turn-on angle	°	− 5
θ_{off}	Turn-off angle	°	11

Mode 1: Normal operation

In this mode, the focus should be on the torque, losses, and torque ripple performances. Thus, they are set as the optimization objectives. The objective function in this mode is to maximize the average torque and minimize the loss and torque ripple, which is given as

$$\text{Objectives}: \begin{cases} \max T_{\text{avg\_ 1}} \\ \min P_{\text{loss\_ 1}} \\ \min T_{\text{ripple\_ 1}} \end{cases} \tag{8.1}$$

where $T_{\text{avg\_1}}$, $P_{\text{loss\_1}}$, and $T_{\text{ripple\_1}}$ are the output torque, loss, and torque ripple in Mode 1, respectively.

Mode 2: Start or climbing

The start and stop operation conditions always occur during urban driving, as shown in Fig. 8.2. In the mode of startup operation of the motor, high torque is the foremost requirement. It is set as the operation objective in this mode. The driving cycle of the climbing is similar to the start operation, which also requires high output torque. Thus, the start and climbing modes are combined as one independent driving mode. The objective function in this mode is given as

$$\text{Objective}: \quad \max T_{\text{avg\_ 2}} \tag{8.2}$$

where $T_{\text{avg\_2}}$ is the output torque in Mode 2.

Mode 3: Open circuit fault operation

The torque ripple is the major concern in this mode due to the operation's reliability. Thus, the torque ripple is selected as the optimization objective. The objective function in this mode is given as

$$\text{Objective}: \quad \min T_{\text{ripple\_3}} \tag{8.3}$$

where $T_{\text{ripple\_3}}$ is the torque ripple in Mode 3.

Mode 4: Short circuit fault operation

The most likely scenario in this mode is that two adjacent coils are short-circuited. Thus, this situation will be considered. Particular attention should be paid to the short-circuit condition. The objective function in this mode is given as

$$\text{Objective}: \quad \min T_{\text{ripple\_4}} \tag{8.4}$$

where $T_{\text{ripple\_4}}$ is the torque ripple in Mode 4.

Mode 5: High-speed operation

Since the net loss, including copper loss and core loss, has a major influence on the motor efficiency and the operation temperature in this mode, it has been selected as the objective. The objective function in this mode is given as

$$\text{Objective}: \quad \min P_{\text{loss\_5}} \tag{8.5}$$

where $P_{\text{loss\_5}}$ is the loss in Mode 5.

The angle position control (APC) method will be investigated for all five driving modes since it is the commonly used control method for SRMs [23, 24].

(b) Sensitivity analysis

A comprehensive sensitivity analysis is adopted to simplify the optimization process by selecting and dividing the optimization variables into different subspaces [25]. The design variables are listed in Table 8.2, and the global sensitivity of f with regard to parameter x_i can be achieved.

Figure 8.6 shows the sensitivity analysis results of the two control and six geometry parameters for the three objectives in Mode 1. Besides, the sensitivity results of the other four driving modes are presented in Fig. 8.7. The sensitive variables can be identified from Figs. 8.6 and 8.7. The results are shown in Table 8.3. Besides, representative speeds of each mode are presented in Table 8.3.

One driving mode may have several influential variables, and the same variable may significantly affect the objectives of multiple driving modes. For example, there are four crucial variables in Mode 1, i.e., θ_{on}, θ_{off}, g, and β_s, in which θ_{on} also shows high impact on the objectives of modes 2, 3 and 5. The crossover of the driving modes and their influential variables highly increases the difficulty of the optimization since different variables should be optimized for the driving modes. Taking variables θ_{off} and g as an example, θ_{off} is important for Modes 1, 2, and 5, while g significantly impacts the objectives of Modes 1, 3, and 4. Thus, the selection of the optimal values of θ_{off} and g cannot be determined by the same driving modes. Thus, analyzing the

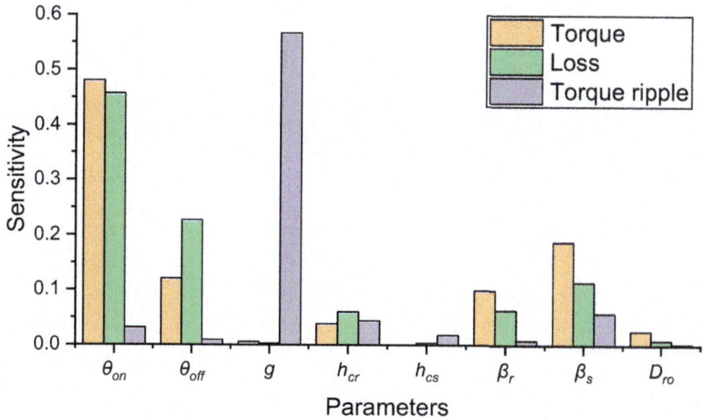

Fig. 8.6 Sensitivity indices for Mode 1

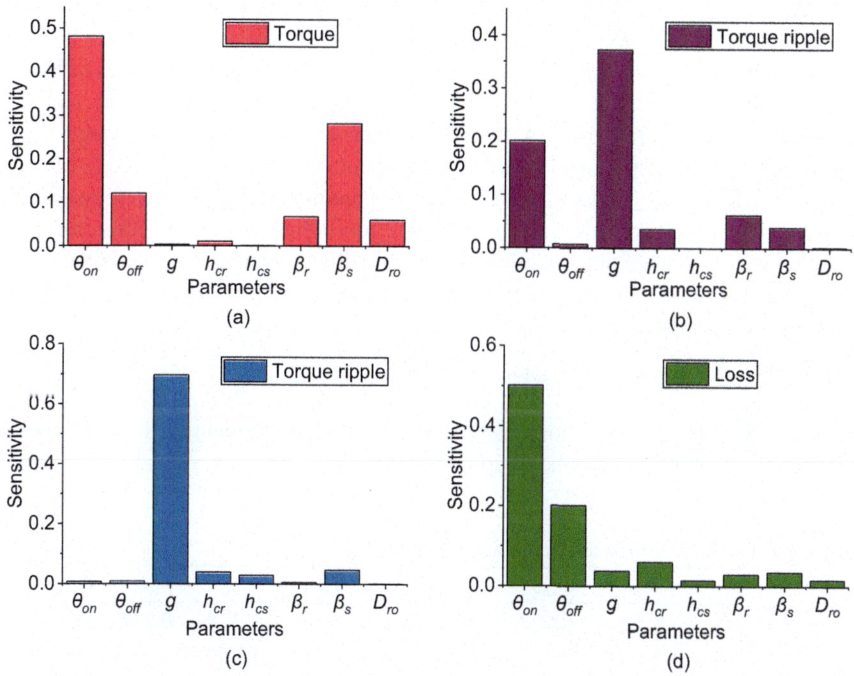

Fig. 8.7 Sensitivity indices **a**–**d** for modes 2–5 respectively

correlation of all the driving modes is necessary to simplify the optimization process further.

Table 8.3 Selection of sensitive variables

Driving modes	Sensitive variables	Representative speed
Mode 1: Normal operation	θ_{on}, θ_{off}, g, β_s	$n = n_{rated}$
Mode 2: Start or climbing	θ_{on}, θ_{off}, β_s	$n \leq 0.5n_{rated}$
Mode 3: Open circuit fault operation	θ_{on}, g	$n = n_{rated}$
Mode 4: Short circuit fault operation	g	$n = n_{rated}$
Mode 5: High-speed operation	θ_{on}, θ_{off}	$n = 1.5–2n_{rated}$

(c) Correlation analysis

The Pearson correlation coefficients [26] between different optimization objectives in the five driving modes are calculated to show the relationship between any two objectives. The value close to 1 indicates a high correlation. As shown in Table 8.4, T_{avg_1} and T_{avg_2} are positively correlated since the Pearson correlation coefficient between the two objectives is 0.992, which is very close to 1. The same relationship can be found between P_{loss_1} and P_{loss_5} and T_{ripple_1} and T_{ripple_4}, where the Pearson correlation coefficients are 0.971 and 0.989, respectively. Thus, the analysis of T_{avg_2}, P_{loss_5}, and T_{ripple_4} can be substituted by T_{avg_1}, P_{loss_5} and T_{ripple_1}, respectively. It should be noted that in Table 8.4, the Pearson coefficient between T_{avg_1} and P_{loss_1} is 0.975. This correlation is similar to that between T_{avg_1} and P_{loss_5}, T_{avg_1} and P_{loss_1}, and T_{avg_2} and P_{loss_5}. However, the optimization goal aims to maximize torque while minimizing loss. This indicates conflicting objectives. Therefore, the analysis of loss-related objectives cannot be substituted by torque-related objectives. After the correlation analysis, only four objectives in two modes are selected for the optimization, as listed in Table 8.5.

(d) Sequential Taguchi method

Figure 8.8 illustrates the flowchart of the multilevel sequential Taguchi robust optimization method. It can be divided into six steps.

Table 8.4 Correlation analysis of optimization objectives

	T_{avg_1}	P_{loss_1}	T_{ripple_1}	T_{avg_2}	T_{ripple_3}	T_{ripple_4}	P_{loss_5}
T_{avg_1}	1	*0.975*	0.215	0.992	0.368	0.128	*0.939*
P_{loss_1}	*0.975*	1	0.143	*0.960*	0.313	0.057	0.971
T_{ripple_1}	0.215	0.143	1	0.218	0.707	0.989	0.215
T_{avg_2}	0.992	*0.960*	0.218	1	0.373	0.134	*0.936*
T_{ripple_3}	0.368	0.313	0.707	0.373	1	0.661	0.355
T_{ripple_4}	0.128	0.057	0.989	0.134	0.661	1	0.135
P_{loss_5}	*0.939*	0.971	0.215	*0.936*	0.355	0.135	1

Italicized numbers indicate that the two objectives are highly correlated but have opposite optimization directions.

Table 8.5 Selection of optimization mode and objectives

Objectives	Variables	Driving mode
T_{avg_1}	θ_{on}, θ_{off}, β_s	Mode 1
P_{loss_1}	θ_{on}, θ_{off}, β_s	Mode 1
T_{ripple_1}	g	Mode 1
T_{ripple_3}	θ_{on}, g	Mode 3

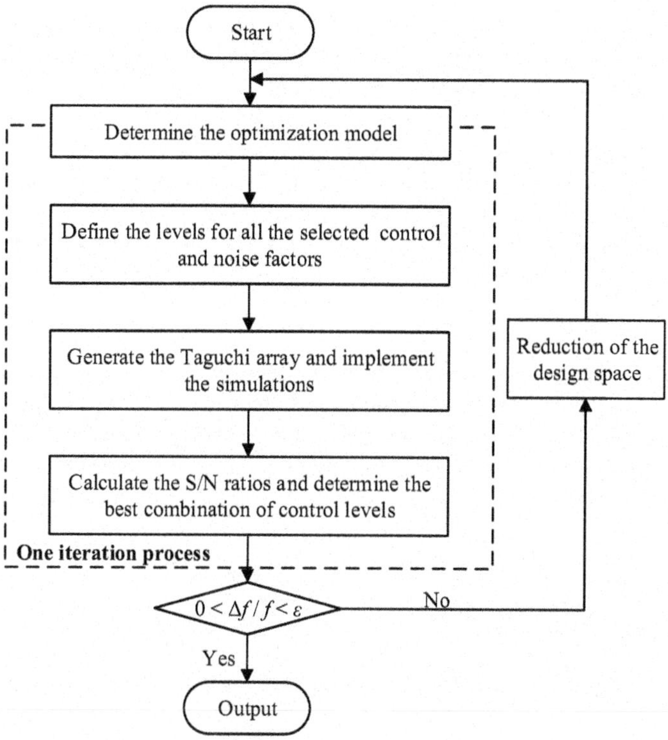

Fig. 8.8 Flowchart of the multilevel sequential Taguchi robust optimization method

Step 1: Determine the optimization model combining all the optimization objectives in the driving modes. The optimization model is defined as

$$\begin{aligned}
\min : f(\mathbf{x}_s) &= w_1 \frac{(T_{avg_1})_{initial}}{T_{avg_1}} + w_2 \frac{P_{loss_1}}{(P_{loss_1})_{initial}} \\
&\quad + w_3 \frac{T_{ripple_1}}{(T_{ripple_1})_{initial}} + w_4 \frac{T_{ripple_3}}{(T_{ripple_3})_{initial}} \\
\text{s.t.} \quad & g_1(\mathbf{x}_s) = 0.85 - \eta \le 0 \\
& g_2(\mathbf{x}_s) = sf - 0.6 \le 0 \\
& g_3(\mathbf{x}_s) = J_c - 6.5 \le 0 \\
& \mathbf{x}_{sl} \le \mathbf{x}_s \le \mathbf{x}_{su}
\end{aligned} \tag{8.6}$$

Table 8.6 Optimization variables and ranges

Parameter	Unit	Ranges
θ_{on}	°	[−6, −4]
θ_{off}	°	[9, 11]
β_s	°	[10, 14]
g	mm	[0.5, 1]

where $\mathbf{x}_s, f, g_i$ are the design parameter vector, objective, and constraints of motor, respectively; $\mathbf{x}_s$ consists of motor and control parameters; $\mathbf{x}_{sl}$ and $\mathbf{x}_{su}$ are the lower boundary and upper boundary of $\mathbf{x}_s$, respectively; η, sf and J_c (in A/mm^2) represent the efficiency, slot fill factor and current density, respectively, and w_1, w_2, w_3, and w_4 are the weighting factors. In this example, since T_{avg_1}, P_{loss_1}, and T_{ripple_1} are the representation of T_{avg_2}, P_{loss_5}, and T_{ripple_4}, respectively, the values of w_1, w_2, w_3, and w_4 are assigned as 2, 2, 2, and 1, respectively. The next several steps will constitute one iteration process, which is the main difference between the proposed sequential Taguchi method and the conventional Taguchi method.

Step 2: Define the levels for all the selected optimization variables and noise factors

Table 8.6 lists the optimization variables and their ranges. Table 8.7 lists the design levels of optimization variables for each iteration process. As shown, there are three levels for each optimization variable. The data in column Iteration 1 in Table 8.7 represents the initial levels of each optimization variable. The following steps will introduce the selection of the levels in the other iteration process. The air gap g is selected as the only noise factor, and its design levels are set as − 0.1 mm and + 0.1 mm, considering the outer diameter size.

Table 8.7 Levels of optimization variables

Variable	Level	Iteration 1	Iteration 2	Iteration 3	Iteration 4
θ_{on} (°)	1	− 6	− 5	− 4.5	− 4.25
	2	− 5	− 4.5	− 4.25	− 4.13
	3	**− 4**	**− 4**	**− 4**	− 4
θ_{off} (°)	1	**9**	9	9.5	9.75
	2	10	9.5	9.75	9.88
	3	11	**10**	**10**	10
β_s (°)	1	**10**	10	10.5	10.75
	2	12	**11**	**11**	11
	3	14	12	11.5	11.25
g (mm)	1	0.5	0.75	0.88	0.94
	2	0.75	0.88	0.94	0.97
	3	**1**	**1**	**1**	1

The best values of each parameter are marked in bold

Step 3: Generate the orthogonal array based on the design levels of the control and noise factors and implement the simulations according to the established array

Table 8.8 lists the orthogonal array generated from these factors. As shown, it has nine rows defined by the optimization variables. 1, 2, and 3 in the table represent the indices of the corresponding levels presented in Table 8.7. During one iteration, 18 (9 × 2) combinations of FEM simulations will be performed to evaluate the performances in mode 1, and 6 (3 × 2) combinations of FEM analyses will be considered to determine the performances in mode 3. More samples are analyzed for Mode 1 than for Mode 3 because there are four variables in Mode 1 while only two in Mode 3, as shown in Table 8.5.

Step 4: Calculate the signal-to-noise (S/N) ratios and determine the best combination of design parameter levels for the optimization

The calculation of the S/N ratios can be divided into two main steps: The first is to compute the S/N ratio for each row of the orthogonal array, and the respective equation is given by

$$SN(i) = -10 \times \lg\left[\frac{1}{2} \times \sum_{j=1}^{2} F^2(i,j)\right] \tag{8.7}$$

where i (1, 2, …, 9) and j (1, 2, 3, 4) are the row number in Table 8.8 and the index of the noise factor (variable), respectively.

Then, the average S/N ratios for all levels of optimization variables based on the S/N ratios of each row are computed. For example, the average S/N ratio for the first level of θ_{on} is the average value of $SN(6)$, $SN(7)$, and $SN(9)$. The combination of the best design levels for the optimization variables can be determined by the S/N ratios. The higher S/N ratio means a smaller F. Thus, the level of each optimization variable, which reveals the highest S/N ratio, is selected to determine the overall best design.

Table 8.8 Orthogonal array

No.	θ_{on}	θ_{off}	β_s	g
1	3	2	3	1
2	3	3	1	2
3	2	1	3	2
4	2	3	2	1
5	2	2	1	3
6	1	3	3	3
7	1	1	1	1
8	3	1	2	3
9	1	2	2	2

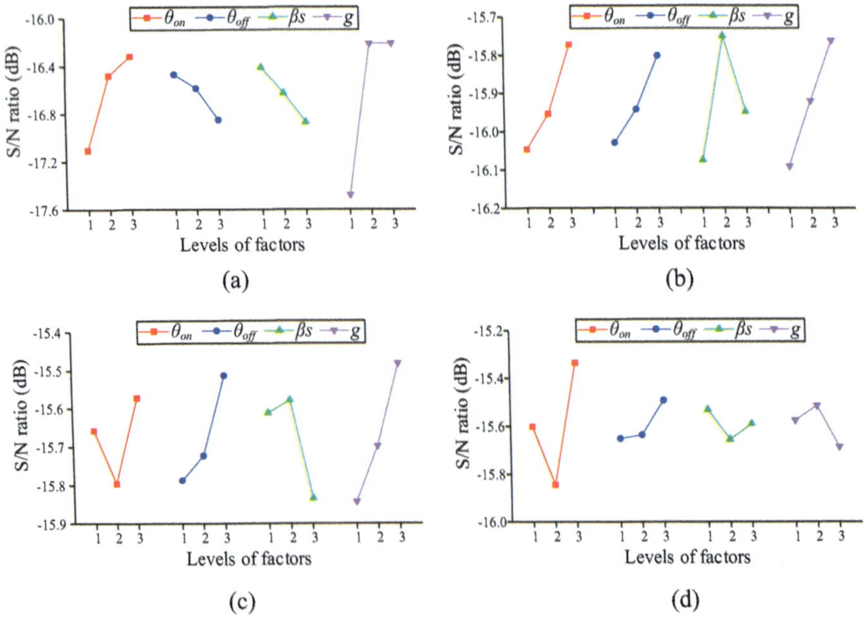

Fig. 8.9 S/N ratios. **a** Iteration 1, **b** iteration 2, **c** iteration 3, and **d** iteration 4

Figure 8.9 illustrates the S/N ratios. As shown in Fig. 8.9a, levels 3, 1, 1, and 3 are selected for θ_{on}, θ_{off}, g, and β_s, respectively, which correspond to − 4, 9, 10, 1 in the column Iteration 1 in Table 8.7.

Step 5: Calculate the value of the selection criterion of the selected combination of design levels

It should be noted that Steps 2–5 are considered as one iteration process. The idea for selecting a sequential iteration process is to improve the performance further and ensure that the optimization process is converged and that the performance of the identified optimal solution cannot be further improved.

Step 6: Termination judgment

Compare the values of F between two successive iterations to judge whether the optimal solution meets the convergence condition. The value of F in iteration 1 should be compared with that of the initial design. If their relative error is larger than 0 and smaller than a given value ε, terminate the iteration progress and output the optimal solution. Otherwise, reduce the design space and go to Step 2 to perform another iteration. If the relative error is completely equal to 0, the values for each selected control factor are the same. In this example, the value of ε is set as 3%. The best combination of optimization variables in each iteration is listed in Table 8.9, as well as the values of the objectives in each driving mode. As shown in Fig. 8.9b, c, the same levels are selected in Iterations 2 and 3, i.e., the best combinations of the optimization variables in Iterations 2 and 3 are the same. The sequential Taguchi method is terminated after four iterations. Since F of Iterations 2 and 3 achieves the lowest value, the optimal design is achieved and listed in Table 8.9.

Figure 8.10 illustrates the flux density of the initial design and the optimal design at the aligned position when the two poles of one phase are excited with 50 A. Generally, the optimal design exhibits lower flux density than the initial design. The initial design is more likely to be saturated, and the optimal design holds a stronger ability to produce output torque. Table 8.9 shows that the optimal solution achieved by the proposed optimization method can significantly reduce the torque ripples in Modes 1, 3, and 4. Compared with the initial design, the torque ripples in Modes 1, 3, and 4 can be reduced by about 60%, 18%, and 60%, respectively. Besides, although the torque in Mode 1 is reduced by about 7%, the losses also decrease by about 12%. As the definition of F is to evaluate the comprehensive performance of the SRM drive system considering multiple driving cycles, the comprehensive performance is improved by about 21%.

Moreover, the torque curves of the initial design and optimal design in Modes 1 and 3 are presented in Figs. 8.11 and 8.12, respectively. They illustrate the changes of torque and torque ripple more intuitively. Compared with Mode 1, the minimum torque value in Mode 3 has been greatly reduced since the open phase loses the ability to generate the force pulling the rotor from the unaligned position to the aligned position within the range of the conduction angle.

Table 8.9 Performance comparison

Parameter	Unit	Initial	Iteration 1	Iterations 2/3	Iteration 4
θ_{on}	°	−5	−4	−4	−4
θ_{off}	°	11	9	10	10
β_s	°	13.5	10	11	10.75
g	mm	0.5	1	1	0.97
T_{avg_1}	N m	126.72	115.46	117.60	121.03
P_{loss_1}	kW	3.04	2.63	2.67	2.71
T_{ripple_1}	%	83.84	74.35	32.70	39.74
T_{avg_2}	N m	462.58	420.89	449.65	460.43
T_{ripple_3}	%	170.32	162.66	138.72	142.57
T_{ripple_4}	%	83.08	75.60	33.27	40.30
P_{loss_5}	kW	2.44	1.83	1.92	1.94
f	–	7	6.65	5.51	5.66

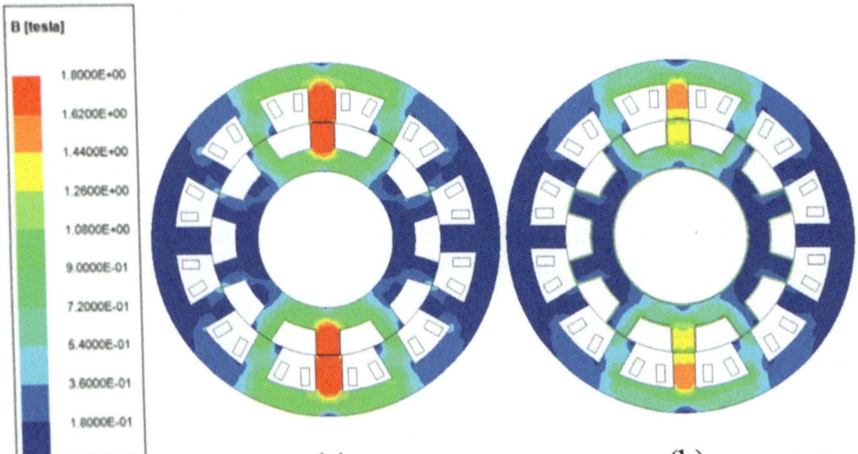

Fig. 8.10 Flux density at the aligned position when the two stator poles of one phase are excited with 50 A. **a** Initial design and **b** optimal design

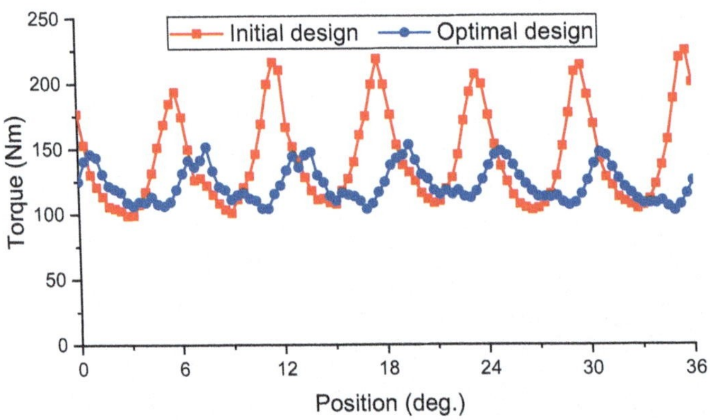

Fig. 8.11 Comparison of torque in mode 1

8.4 Summary

This chapter presents an application-oriented design optimization method for SRMs. To illustrate this method, a novel system-level robust design optimization considering multiple driving modes is proposed for a 12/10 SRM drive system for EV application. Five driving modes of the motor system are defined according to the driving cycle. To improve the optimization efficiency, a sensitivity analysis is implemented to identify influential variables as the design parameters for the optimization. A correlation analysis is implemented to determine the coherence of the objective functions of all

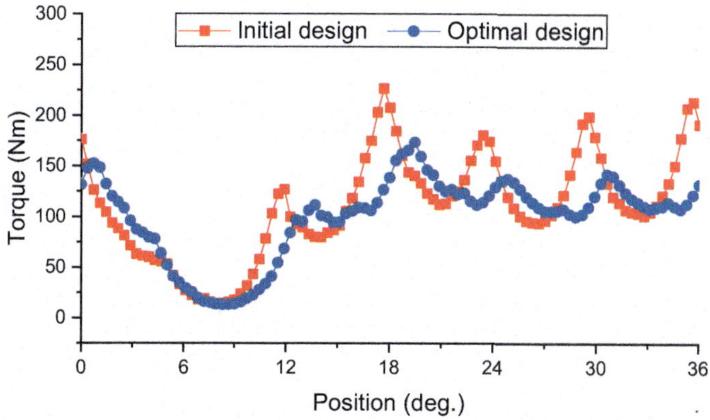

Fig. 8.12 Comparison of torque in mode 3

driving modes to simplify the optimization process. A sequential Taguchi method is proposed and applied to find the optimal solution.

References

1. Li SF, Zhang S, Habetler TG, Harley RG. Modeling, design optimization, and applications of switched reluctance machines-a review. IEEE Trans Ind Appl. 2019;55(3):2660–81. https://doi.org/10.1109/Tia.2019.2897965.
2. Duan Y, Ionel DM. A review of recent developments in electrical machine design optimization methods with a permanent-magnet synchronous motor benchmark study. IEEE Trans Ind Appl. 2013;49(3):1268–75. https://doi.org/10.1109/tia.2013.2252597.
3. Lei G, Bramerdorfer G, Liu C, Guo Y, Zhu J. Robust design optimization of electrical machines: a comparative study and space reduction strategy. IEEE Trans Energy Convers. 2021;36(1):300–13. https://doi.org/10.1109/TEC.2020.2999482.
4. Ma B, Lei G, Zhu JG, Guo YG, Liu CC. Application-oriented robust design optimization method for batch production of permanent-magnet motors. IEEE Trans Ind Electron. 2018;65(2):1728–39. https://doi.org/10.1109/tie.2017.2748046.
5. Lei G, Li YB, Liu CC, Ma B, Guo YG, Zhu JG, et al. Application-oriented design optimization of flux-switching PM motors. In: 20th IEEE International conference on electrical machines and systems (ICEMS), Sydney, Australia; 2017.
6. Diao K, Sun X, Lei G, Guo Y, Zhu J. Multimode optimization of switched reluctance machines in hybrid electric vehicles. IEEE Trans Energy Convers. 2021;36(3):2217–26. https://doi.org/10.1109/tec.2020.3046721.
7. Diao K, Sun X, Lei G, Guo Y, Zhu J. Application-oriented system level optimization method for switched reluctance motor drive systems. In: IEEE 9th International power electronics and motion control conference (IPEMC-ECCE Asia), Nanjing, Peoples Republic of China; 2020. p. 472–7.
8. Yu Z, Gan C, Ni K, Chen Y, Qu R. Dual-electric-port bidirectional flux-modulated switched reluctance machine drive with multiple charging functions for electric vehicle applications. IEEE Trans Power Electron. 2021;36(5):5818–31. https://doi.org/10.1109/tpel.2020.3029822.

9. Yan W, Chen H, Liao S, Liu Y, Cheng H. Design of a low-ripple double-modular-stator switched reluctance machine for electric vehicle applications. IEEE Trans Transp Electrification. 2021;7(3):1349–58. https://doi.org/10.1109/tte.2021.3057133.
10. Diao K, Sun X, Lei G, Bramerdorfer G, Guo Y, Zhu J. System-level robust design optimization of a switched reluctance motor drive system considering multiple driving cycles. IEEE Trans Energy Convers. 2021;36(1):348–57. https://doi.org/10.1109/tec.2020.3009408.
11. Lei G, Guo Y, Zhu J. Multidisciplinary design optimization methods for electrical machines and drive systems. Berlin, Heidelberg: Springer; 2016.
12. Diao K, Sun X, Lei G, Bramerdorfer G, Guo Y, Zhu J. Robust design optimization of switched reluctance motor drive systems based on system-level sequential Taguchi method. IEEE Trans Energy Convers. 2021;36(4):3199–207. https://doi.org/10.1109/tec.2021.3085668.
13. Bramerdorfer G, Tapia JA, Pyrhonen JJ, Cavagnino A. Modern electrical machine design optimization: techniques, trends, and best practices. IEEE Trans Ind Electron. 2018;65(10):7672–84. https://doi.org/10.1109/tie.2018.2801805.
14. Xue XD, Cheng KWE, Ng TW, Cheung NC. Multi-objective optimization design of in-wheel switched reluctance motors in electric vehicles. IEEE Trans Ind Electron. 2010;57(9):2980–7. https://doi.org/10.1109/TIE.2010.2051390.
15. Zhu X, Xiang Z, Quan L, Wu W, Du Y. Multimode optimization design methodology for a flux-controllable stator permanent magnet memory motor considering driving cycles. IEEE Trans Ind Electron. 2018;65(7):5353–66.
16. Li GJ, Ojeda J, Hoang E, Lecrivain M, Gabsi M. Comparative studies between classical and mutually coupled switched reluctance motors using thermal-electromagnetic analysis for driving cycles. IEEE Trans Magn. 2011;47(4):839–47. https://doi.org/10.1109/TMAG.2011.2104968.
17. Chen L, Xu H, Sun X. A novel strategy of control performance improvement for six-phase permanent magnet synchronous hub motor drives of EVs under new European driving cycle. IEEE Trans Veh Technol. 2021;70(6):5628–37. https://doi.org/10.1109/TVT.2021.3079576.
18. Lazari P, Wang J, Chen L. A computationally efficient design technique for electric-vehicle traction machines. IEEE Trans Ind Appl. 2014;50(5):3203–13.
19. Yang Z, Shang F, Brown IP, Krishnamurthy M. Comparative study of interior permanent magnet, induction, and switched reluctance motor drives for EV and HEV applications. IEEE Trans Transp Electrification. 2015;1(3):245–54. https://doi.org/10.1109/TTE.2015.2470092.
20. Degano M, Carraro E, Bianchi N. Selection criteria and robust optimization of a traction PM-assisted synchronous reluctance motor. IEEE Trans Ind Appl. 2015;51(6):4383–91.
21. Diao K, Bramerdorfer G, Sun X, Yang Z, Han S. Multiobjective design optimization of a novel switched reluctance motor with unequal alternating stator yoke segments. IEEE Trans Transp Electrification. 2023;9(1):512–21. https://doi.org/10.1109/tte.2022.3174337.
22. Diao K, Sun X, Bramerdorfer G, Cai Y, Lei G, Chen L. Design optimization of switched reluctance machines for performance and reliability enhancements: a review. Renew Sust Energy Rev. 2022;168. https://doi.org/10.1016/j.rser.2022.112785
23. Li S, Zhang S, Habetler TG, Harley RG. Modeling, design optimization, and applications of switched reluctance machines—a review. IEEE Trans Ind Appl. 2019;55(3):2660–81. https://doi.org/10.1109/TIA.2019.2897965.
24. Ding W, Yang S, Hu Y. Performance improvement for segmented-stator hybrid-excitation SRM drives using an improved asymmetric half-bridge converter. IEEE Trans Ind Electron. 2019;66(2):898–909. https://doi.org/10.1109/TIE.2018.2833034.
25. Zhu X, Huang J, Quan L, Xiang Z, Shi B. Comprehensive sensitivity analysis and multiobjective optimization research of permanent magnet flux-intensifying motors. IEEE Trans Ind Electron. 2019;66(4):2613–27. https://doi.org/10.1109/TIE.2018.2849961.
26. Sun X, Shi Z, Lei G, Guo Y, Zhu J. Multi-objective design optimization of an IPMSM Based on multilevel strategy. IEEE Trans Ind Electron. 2020.

Chapter 9
Conclusions and Future Trends

Abstract This chapter concludes the book with two sections: conclusions and future trends. The first section briefly summarizes the main analysis, design, and optimization methods for switched reluctance motors (SRMs) and drive systems. The second section provides some discussions and suggestions for future work in this field, including the application of machine learning.

Keywords Application-oriented design optimization · Electric vehicles · Multi-disciplinary optimization · Robust topology optimization · Transportation

9.1 Conclusions

In this book, the following design fundamentals and optimization methods have been investigated for switched reluctance motors (SRMs) and drive systems, including:

(1) Novel design topologies. A set of novel topologies of SRMs have been derived from the conventional structure from the perspective of increasing the utilization of the magnetic performance, reducing the ripple, or adapting to the specific applications. SRMs with segmented/double stator/rotor, hybrid excitation, skewed teeth, axial flux, linear structure, and their combinations have been presented. The segmented structure can improve the output torque and fault-tolerance ability but cause high manufacturing complexity and potentially mechanical weakness. The hybrid-excitation SRM can further enhance the power due to the extra flux path provided by the permanent magnets. The main benefit of the double stator/rotor and skewed-teeth topologies is that they reduce vibration. Moreover, axial-flux SRM and linear SRM are suitable for applications with a high diameter-to-axial-length ratio and vertical actuation systems, respectively.

(2) Nonlinear magnetic parameter design method. The nonlinear inductance/flux linkage versus position is the inherent feature of SRMs. A novel magnetic parameter design method was proposed based on this feature, and corresponding equations of the main dimensions are deduced from the co-energy expressed by the

X. Sun et al., *Multi-objective Design Optimization of Switched Reluctance Motor Drive Systems*, https://doi.org/10.1007/978-981-96-0672-6_9

area on the flux-current figure. The proposed design methodology can provide satisfactory calculation of the design variables and reduce the over-dependence on the designer's experience.

(3) System-level deterministic optimization. The multi-objective system-level optimization method based on a multi-level strategy, surrogate model, and genetic algorithm is proposed for SRMs. The design variables in motor and control parts have been optimized at the system level rather than the component level to achieve the system's best performance. A segmented-rotor SRM, along with the angle position control method, has been investigated as an example. The proposed method exhibits many benefits for the drive system, including high efficiency and low torque ripple.

(4) System-level robust optimization. The main concern in mass production is the influence of uncertainties/tolerances. A robust optimization method based on the sequential Taguchi method is proposed for SRMs. Two examples are investigated to compare component-level and system-level robust approaches and system-level deterministic and robust approaches. The former comparison shows that, compared with the component-level Taguchi method, the system-level Taguchi method can achieve a higher level of robustness with better output performance. The latter comparison shows that the reliability of the proposed system-level robust approach can be greatly improved.

(5) Application-oriented optimization. The optimization process's ultimate target is finding the best solutions to meet the specific application requirements. Various design optimization techniques can be utilized, including deterministic and robust approaches. Two SRM examples are presented, one for the belt-driven generator in hybrid electric vehicles and the other for the EV drive, to show the efficiency and necessity of the proposed method.

9.2 Future Trends

The previous chapters focused on the design methods, novel topologies, and optimization techniques. As these sections revealed, the research area of design optimization techniques for SRMs is widespread. Since PM motors occupy most of the market share, most scholars have not emphasized SRMs [1, 2]. Therefore, potential prospective research topics regarding the design optimization of SRMs exhibit great potential significance [3, 4]. Several future trends based on the existing research experience are presented as follows.

9.2.1 Robust Topology Optimization of SRMs

Topology optimization aims to obtain the optimal material layout in the prescribed design domain (like rotor or stator) to obtain the best structural performance by transforming the design optimization problem into a material distribution problem [5, 6]. However, conventional topology optimization mainly focuses on finding designs that perform well under idealized conditions, assuming perfect material properties and deterministic loading conditions, and cannot handle uncertainties [7–9]. Robust topology optimization considers these uncertainties and seeks to find designs that are less sensitive to variations, ensuring improved performance and reliability.

Robust topology optimization is an emerging topic that would greatly suit SRMs to reduce material costs further, obtain a faster dynamic response, and improve cooling conditions. SRMs exhibit inherently better mechanical robustness compared with motors with PMs [10]. Thus, the topology design domain of SRMs is more flexible, and more solutions can be carried out to remove the abundant materials in the stator and rotor under performance fluctuation.

9.2.2 *System-Level Multi-disciplinary Design Optimization of SRMs*

In conventional design optimization works, most emphases were placed on the magnetic behaviors of SRMs, while the performances in other fields like temperature and stress are normally ignored. As discussed before, the life of windings and lamination materials is determined by the rise in temperature, and deformation will occur due to high stress [11]. The coupling relationships between electromagnetic, thermal, and mechanical fields should be evaluated to fully use the materials and ensure that the final design's performance is within the constraints. The stable operation of SRMs is limited by the temperature rise of the power converter, and the device performance will deteriorate if the components in the power converter exceed the reasonable domain [12].

System-level-based multi-disciplinary design optimization is an approach to engineering design that considers the interactions and dependencies among multiple disciplines or subsystems in a complex system. It involves optimizing the design of the entire system rather than optimizing individual components in isolation. System-level-based multi-disciplinary design optimization seeks to address this limitation by integrating the various disciplines and optimizing the system as a whole. It considers the interdependencies, trade-offs, and synergies between different subsystems to achieve an optimal design that satisfies multiple performance criteria.

9.2.3 *Application-Oriented Design Optimization of SRMs for Future Transportation*

A great amount of research efforts on SRMs have been shifted to the application in EVs and HEVs [13–17]. The literature shows that SRMs could perform similarly to PM motors [2]. Load test results over the entire speed range, which verify the predictions, were presented in [18]. However, to the authors' best knowledge, few SRM drives have been installed on commercial EVs/HEVs except for prototypes.

Another field of future transportation is the more electric aircraft (MEA). In this scenario, SRMs outperform PM motors and induction motors due to their superior thermal and mechanical robustness without PMs and coils in the rotor [19]. The SRM applied in aircraft systems mainly acts as the starter/generator, in which the SRM first provides the torque to spin the engine to the idling speed and then works in generator mode for power generation.

9.2.4 Application of AI Technology to the Analysis and Design Optimization of SRMs and Drive Systems

The application of AI technology for the design optimization of SRMs and drive systems holds significant potential for advancing performance, efficiency, and sustainability in various industrial and transportation applications.

AI-driven design optimization could efficiently explore the design space by considering conflicting objectives, uncertainties, and variability in design parameters and generating innovative design topologies using generative design algorithms. For material selection and integration, machine learning algorithms aid in recommending optimal material choices for SRM components and drive system parts, considering material properties, manufacturing constraints, and performance requirements. Regarding manufacturing, AI-driven design optimization exploits the design freedom afforded by additive manufacturing processes to create complex geometries and customized components, achieving optimal performance and manufacturability. Besides, it is also beneficial to sustainability through life cycle assessment and optimization for ease of maintenance, repairability, and recyclability.

9.2.5 Application of 3D Printing Technology to the Development of SRMs

3D printing enables the creation of complex geometries that are difficult or impossible to achieve with conventional manufacturing methods. This allows for the customization and optimization of SRM components to improve performance, efficiency, and functionality. These components can be designed with lightweight structures facilitated by intricate internal geometries achievable only through 3D printing, resulting in enhanced energy efficiency and seamless integration into various systems. Additionally, on-demand manufacturing capabilities offered by 3D printing reduce costs, minimize waste, and provide flexibility in meeting fluctuating demand. Advancements in 3D printing materials further enhance the mechanical, thermal, and electrical properties of SRM components, bolstering overall performance and reliability.

Future developments in multi-material 3D printing hold promise for fabricating electrical machine components with integrated functionalities, leading to more compact and efficient designs.

References

1. Shi Z, Sun X, Lei G, Yang Z, Guo Y, Zhu J. Analysis and optimization of radial force of permanent-magnet synchronous hub motors. IEEE Trans Magn. 2020;56(2):1–4. https://doi.org/10.1109/TMAG.2019.2953731.
2. Kiyota K, Chiba A. Design of switched reluctance motor competitive to 60-kW IPMSM in third-generation hybrid electric vehicle. IEEE Trans Ind Appl. 2012;48(6):2303–9. https://doi.org/10.1109/TIA.2012.2227091.
3. Diao K, Sun X, Bramerdorfer G, Cai Y, Lei G, Chen L. Design optimization of switched reluctance machines for performance and reliability enhancements: a review. Renew Sust Energy Rev. 2022;168. https://doi.org/10.1016/j.rser.2022.112785
4. Feng L, Sun X, Bramerdorfer G, Zhu Z, Cai Y, Diao K, et al. A review on control techniques of switched reluctance motors for performance improvement. Renew Sust Energ Rev. 2024;199:114454. https://doi.org/10.1016/j.rser.2024.114454.
5. Zhao Q, Chen X, Ma Z-D, Lin Y. Robust topology optimization based on stochastic collocation methods under loading uncertainties. Math Probl Eng. 2015;2015. https://doi.org/10.1155/2015/580980
6. Xia M, Sun D, Yang S. A novel methodology for robust topology optimization considering manufacturing errors and topology deviations. IEEE Trans Magn. 2022;58(9):1–4. https://doi.org/10.1109/TMAG.2022.3172854.
7. Lei G, Bramerdorfer G, Ma B, Guo Y, Zhu J. Robust design optimization of electrical machines: multi-objective approach. IEEE Trans Energy Convers. 2021;36(1):390–401. https://doi.org/10.1109/TEC.2020.3003050.
8. Diao K, Sun X, Lei G, Bramerdorfer G, Guo Y, Zhu J. Robust design optimization of switched reluctance motor drive systems based on system-level sequential Taguchi method. IEEE Trans Energy Convers. 2021;36(4):3199–207. https://doi.org/10.1109/tec.2021.3085668.
9. Diao K, Sun X, Lei G, Bramerdorfer G, Guo Y, Zhu J. System-level robust design optimization of a switched reluctance motor drive system considering multiple driving cycles. IEEE Trans Energy Convers. 2021;36(1):348–57. https://doi.org/10.1109/tec.2020.3009408.
10. Feng L, Sun X, Guo D, Yao M, Diao K. Advanced torque sharing function strategy with sliding mode control for switched reluctance motors. IEEE Trans Transp Electrification. 2024;10(1):2302–11. https://doi.org/10.1109/TTE.2023.3287159.
11. Lei G, Guo Y, Zhu J. Multidisciplinary design optimization methods for electrical machines and drive systems. Berlin, Heidelberg: Springer; 2016.
12. Chen H, Xu Y. Electromagnetic field analysis coupled model of fluid–structure–thermal simulation of power converter for switched reluctance machine. IEEE Trans Appl Supercond. 2016;26(4):1–6. https://doi.org/10.1109/TASC.2016.2543604.
13. Sun X, Shen Y, Wang S, Lei G, Yang Z, Han S. Core losses analysis of a novel 16/10 segmented rotor switched reluctance BSG motor for HEVs using nonlinear lumped parameter equivalent circuit model. IEEE/ASME Trans Mechatron. 2018;23(2):747–57. https://doi.org/10.1109/TMECH.2018.2803148.
14. Cheng H, Wang L, Xu L, Ge X, Yang S. An integrated electrified powertrain topology with SRG and SRM for plug-in hybrid electrical vehicle. IEEE Trans Ind Electron. 2020;67(10):8231–41. https://doi.org/10.1109/TIE.2019.2947854.

15. Nguyen D, Bahri I, Krebs G, Berthelot E, Marchand C, Ralev I, et al. Efficiency improvement by the intermittent control for switched reluctance machine in automotive application. IEEE Trans Ind Appl. 2019;55(4):4167–82. https://doi.org/10.1109/TIA.2019.2906860.
16. Sun X, Diao K, Yang Z. Performance improvement of a switched reluctance machine with segmental rotors for hybrid electric vehicles. Comput Electr Eng. 2019;77:244–59. https://doi.org/10.1016/j.compeleceng.2019.06.002.
17. Sun X, Wu J, Wang S, Diao K, Yang Z. Analysis of torque ripple and fault-tolerant capability for a 16/10 segmented switched reluctance motor in HEV applications. Compel-Int J Comp Math Electr Electron Eng. 2019;38(6):1725–37. https://doi.org/10.1108/compel-11-2018-0477.
18. Kiyota K, Kakishima T, Chiba A. Comparison of test result and design stage prediction of switched reluctance motor competitive with 60-kW rare-earth PM motor. IEEE Trans Ind Electron. 2014;61(10):5712–21. https://doi.org/10.1109/TIE.2014.2304705.
19. Li S, Zhang S, Habetler TG, Harley RG. Modeling, design optimization, and applications of switched reluctance machines—a review. IEEE Trans Ind Appl. 2019;55(3):2660–81. https://doi.org/10.1109/TIA.2019.2897965.

SPRINGER NATURE

GPSR Compliance

The European Union's (EU) General Product Safety Regulation (GPSR) is a set of rules that requires consumer products to be safe and our obligations to ensure this.

If you have any concerns about our products, you can contact us on ProductSafety@springernature.com

In case Publisher is established outside the EU, the EU authorized representative is:

Springer Nature Customer Service Center GmbH
Europaplatz 3
69115 Heidelberg, Germany

The manufacturer's authorised representative in the EU is Springer Nature Customer Service Centre GmbH, Europaplatz 3, 69115 Heidelberg, Germany. If you have any concerns regarding our products, please contact ProductSafety@springernature.com

Printed and bound by CPI Group (UK) Ltd, Croydon, CR0 4YY
30/09/2026
02225308-0014